SCIENTISTS, ENGINEERS, AND TRACK-TWO DIPLOMACY

A Half-Century of U.S.-Russian Interacademy Cooperation

Glenn E. Schweitzer

NATIONAL RESEARCH COUNCIL
OF THE NATIONAL ACADEMIES

The views expressed are those of the author and do not represent an official policy of the National Academies.

THE NATIONAL ACADEMIES PRESS
Washington, D.C.
www.nap.edu

THE NATIONAL ACADEMIES PRESS   500 Fifth Street, N.W.   Washington, DC 20001

NOTICE: The project that is the subject of this report was approved by the Governing Board of the National Research Council, whose members are drawn from the councils of the National Academy of Sciences, the National Academy of Engineering, and the Institute of Medicine.

This study was supported by the National Research Council.

International Standard Book Number 0-309-09093-8 (Book)
International Standard Book Number 0-309-52799-6 (PDF)

A limited number of copies of this report are available from:

Development, Security, and Cooperation
National Research Council
KECK 547
500 Fifth Street, N.W.
Washington, D.C. 20001

Additional copies of this report are available from the National Academies Press, 500 Fifth Street, N.W., Lockbox 285, Washington, DC 20055; (800) 624-6242 or (202) 334-3313 (in the Washington metropolitan area); Internet, http://www.nap.edu

THE NATIONAL ACADEMIES
Advisers to the Nation on Science, Engineering, and Medicine

The **National Academy of Sciences** is a private, nonprofit, self-perpetuating society of distinguished scholars engaged in scientific and engineering research, dedicated to the furtherance of science and technology and to their use for the general welfare. Upon the authority of the charter granted to it by the Congress in 1863, the Academy has a mandate that requires it to advise the federal government on scientific and technical matters. Dr. Bruce M. Alberts is president of the National Academy of Sciences.

The **National Academy of Engineering** was established in 1964, under the charter of the National Academy of Sciences, as a parallel organization of outstanding engineers. It is autonomous in its administration and in the selection of its members, sharing with the National Academy of Sciences the responsibility for advising the federal government. The National Academy of Engineering also sponsors engineering programs aimed at meeting national needs, encourages education and research, and recognizes the superior achievements of engineers. Dr. Wm. A. Wulf is president of the National Academy of Engineering.

The **Institute of Medicine** was established in 1970 by the National Academy of Sciences to secure the services of eminent members of appropriate professions in the examination of policy matters pertaining to the health of the public. The Institute acts under the responsibility given to the National Academy of Sciences by its congressional charter to be an adviser to the federal government and, upon its own initiative, to identify issues of medical care, research, and education. Dr. Harvey V. Fineberg is president of the Institute of Medicine.

The **National Research Council** was organized by the National Academy of Sciences in 1916 to associate the broad community of science and technology with the Academy's purposes of furthering knowledge and advising the federal government. Functioning in accordance with general policies determined by the Academy, the Council has become the principal operating agency of both the National Academy of Sciences and the National Academy of Engineering in providing services to the government, the public, and the scientific and engineering communities. The Council is administered jointly by both Academies and the Institute of Medicine. Dr. Bruce M. Alberts and Dr. Wm. A. Wulf are chair and vice chair, respectively, of the National Research Council.

www.national-academies.org

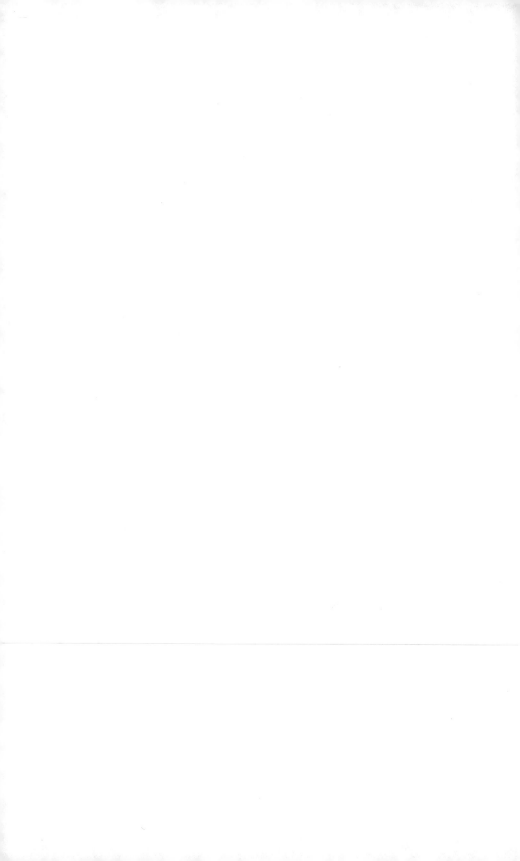

Preface

track-two diplomacy *n* (2003): a term referring to the use of
unofficial channels to facilitate negotiations between governments,
to promote international engagement without arousing hostility,
or to build confidence among elites across international boundaries.[1]

For several centuries, both the United States and Russia have relied on
scientific prowess to help promote the social and economic well-being of
their populations, albeit in very different ways. At the same time, military
science and technology have been crucial in meeting security challenges to
the governments of these countries, from civil wars and from abroad, including the cold war challenges from one another.

During this long period, the governments of the two countries (including the government of the Union of Soviet Socialist Republics) have reached
out to acquire technical expertise from abroad to complement home-grown
scientific capacity. Thus, it is not surprising that achievements in one country
have frequently spread to the other through direct and indirect routes, including exchanges of scientists and students, international trade in high-tech
services and products, joint projects sponsored by the two governments, and
espionage efforts targeting military and industrial secrets.

---

[1] The many definitions of track-two diplomacy all emphasize unofficial channels and
highlight advocacy for peace. This definition, devised by the author, is based on the definitions of *diplomacy* and *elites* in *Merriam-Webster's Collegiate Dictionary,* 11th ed. (Springfield,
Mass.: Merriam-Webster Inc., 2003).

Whatever the past motivations for the bilateral technical contacts of government officials in Washington and in Moscow as well as of specialists at the hundreds of hubs of scientific and technical activities in the two countries, the interests on both sides in staying abreast of the technical achievements of former adversaries remain high. Although Western interest in using Russian technology for space exploration and other cutting-edge endeavors has lessened significantly during the past decade, Russia as a market for Western technologies, Russia as a birthplace of dual-use technologies that might be used irresponsibly by unreliable states, and Russia as an incubator of basic science and new ideas remain of considerable global interest. All the while, Russian specialists continue to look to the United States as the home of many of the world's best technologies, which seem to be essential ingredients of future Russian products if those products are to compete in the global marketplace. In the research arena, Russian specialists know they must keep abreast of achievements in the United States if they are to be meaningful participants in international discussions of modern science.

During the second half of the twentieth century, a formally organized program of scientific cooperation between the U.S. National Academy of Sciences (NAS) and the Soviet Academy of Sciences (ASUSSR), and more recently the Russian Academy of Sciences (RAS), was a major factor in encouraging and facilitating direct cooperation between hundreds of laboratories and thousands of specialists in both Russia and the United States. The academies have organized and managed bilateral programs and convened bilateral and multilateral meetings in the two countries on a wide array of technical and policy topics. In addition, some bilateral pilot programs initiated by the academies have stimulated other organizations to follow with expanded cooperative efforts patterned after the pilot programs. The academies have provided advice to their respective governments, both in reports and in face-to-face meetings with government leaders, and they have often influenced government officials to look favorably at the advantages of scientific engagement over scientific confrontation.

During the 1960s, unprecedented access to Russian specialists, facilities, and data banks was a major motivation for the U.S. government to provide support for the interacademy program. Soon, the acquisition from Russian institutions of technical concepts and information of considerable importance to U.S. scientific endeavors—an effort that built on the newly acquired access—became the primary rationale for continued support. Then, as the Russian economy crumbled, humanitarian concerns and the need to preserve Russian schools of science also became drivers of the cooperative program in Washington.

Although the cooperative programs developed by the National Academies and the Russian Academy of Sciences have the longest history of any U.S.-Russian (including U.S.-Soviet) cooperative programs in science and technology, they have been but a small component of the collaboration of American and Russian scientists and engineers. For example, in the 1960s the Soviet government invited thousands of American scientists to participate in a series of international conferences in Moscow and then to visit research centers in other cities. In the 1970s, the U.S. government took the initiative to establish 11 intergovernmental science and technology agreements, and the National Academy of Sciences played a leading role in only the physics subagreement. Also during the 1970s, U.S.-Russian cooperation in manned space flight increased spectacularly and has continued ever since. By the late 1990s, the U.S. Department of Energy and the Russian Ministry of Atomic Energy had entered into a dozen agreements involving nuclear science and technology, and the U.S. Civilian Research and Development Foundation had come into being with science-oriented programs in Russia that in financial terms and number of participants exceeded the programs of the National Academies severalfold. All the while, the International Research and Exchanges Board (IREX) has for several decades mobilized the U.S. social science community to become engaged in Russia in a major way on a continuing basis as contrasted with the occasional social scientists traveling across the ocean under the auspices of the National Academies. This report does not address these and many other related organized and informal activities, but their contributions to building bridges for science and for peace have been manifold.

From the Russian perspective, as noted earlier, easier access to U.S. technical achievements has always been a big inducement to Russian participation in scientific exchanges and related activities. Since the 1960s, Russian scientists have been eager to gain international recognition for their capabilities, and the interacademy channel provides a prestigious route to this end. In the present economic environment, Russian specialists have been forced to search the globe for any source of financial support to sustain their efforts. The interacademy exchanges help them to establish contacts that sometimes lead to new sources of financing.

Over the decades, the differences in the structures and roles of the National Academies and the Russian Academy of Sciences (as well as the Soviet Academy of Sciences) have caused both confusion and difficulties in organizing cooperative activities. The academies in both countries are honorific organizations. However, the National Academies, with no research facilities to manage, have traditionally emphasized studies that provide advice to the U.S. government on issues that have significant science, engineering, and health dimensions. The RAS has always been primarily concerned with fundamental science investigations carried out at more than 400 research institutes.

In years past, when American participants in cooperative projects expressed interest in working with colleagues outside the academy system, the ASUSSR lost enthusiasm in the projects. At the same time, the ASUSSR had difficulty understanding why the NAS could not simply order American scientists to participate in joint endeavors. In recent years, exchanges have been organized primarily on a scientist-to-scientist basis, thereby alleviating most of the problems surrounding where the scientists work.

Also in earlier times, the Soviet and Russian academies of sciences showed limited interest in cooperation in policy studies. However, during the past few years the RAS has increased its advisory role to government, preparing many dozens of policy documents for government consideration each year. And it has significantly expanded its applied research activities, which often become entwined with science policy issues. At the same time, the RAS has increasingly reached out to involve specialists from other organizations in its international policy-oriented activities. Thus today, the asymmetry in structures and roles of the academies is far less of an impediment to effective cooperation than in the past.

PURPOSE AND SCOPE OF THIS REPORT

Against this background, the purpose of this report is threefold:

1. To provide a brief historical perspective of the evolution of the interacademy program during the past half-century, recognizing that many legacies of the Soviet era continue to influence government approaches in Moscow and Washington and to shape the attitudes of researchers toward bilateral cooperation in both countries. Of special interest is the changing character of the program during the age of *perestroika* (restructuring) in the late 1980s in the Soviet Union.

2. To describe in some detail the significant interacademy activities from late 1991, when the Soviet Union fragmented, to mid-2003. In some instances, the report provides insights into the unusual circumstances that have stimu-

lated program initiatives and into the well-established reasons for promoting international science in the age of globalization. Some of the interacademy activities should be of continuing interest both to government officials and to scientists and engineers in the two countries, because related programs are currently being implemented through other channels with support of the two governments, private foundations, and individual laboratories.

3. To set forth lessons learned about the benefits and limitations of interacademy cooperation and to highlight approaches that have been successful in overcoming difficulties of implementation. These insights should help policy makers both in understanding how scientific cooperation has achieved a special, and seemingly permanent, status in U.S.-Russian relations and in designing future bilateral programs of the academies and other organizations.

This report focuses primarily on the activities organized on the U.S. side by the National Academy of Sciences and implemented by the National Research Council (NRC). Recently, the National Academy of Engineering (NAE) and the Institute of Medicine (IOM) have become more involved in cooperative activities with Russian counterparts, and the importance of this trend for the future is recognized.[2] On the Russian side, most of the projects have involved the active participation of the Russian Academy of Sciences. Even on those occasions that the principal Russian partner has been another organization such as the Ministry of Atomic Energy or the Ministry of Health, the staff of the RAS has provided valuable support for the projects.

This report does not describe the activities of the Russian and U.S. Committees on International Security and Arms Control (CISAC), an interacademy effort that dates back to 1982. These committees have held 30 bilateral meetings, and their activities have been so extensive as to warrant a separate report. At the same time, the complementary nature of the CISAC activities to the programs discussed in this report has been clear for many years—particularly in carrying out track-two diplomacy when intergovernmental diplomatic efforts on sensitive issues such as nuclear confrontation have been at a low ebb.

Thousands of books and scholarly articles have been written about the changes to the environment in Russia that surrounds cooperative programs.

---

[2] The National Academy of Sciences, National Academy of Engineering, Institute of Medicine, and National Research Council are known collectively as the National Academies.

A limited number have addressed the evolution of the Russian scientific establishment (Medvedev, 1978; Holloway, 1983; Vucinich, 1984; Balzer, 1989; Sakharov, 1990; Graham, 1993; Sagdeev, 1994; Marchuk, 1995; Gokhberg, 1997; Josephson, 1997). Only a handful has focused specifically on U.S.-Russian cooperation in science and technology, and some of these works are cited in this report. The single previous report that targeted exclusively interacademy cooperation—the Kaysen report published in 1977—is discussed in Chapter 2.

Much of the information set forth in this report is not available elsewhere. The National Academies Press (NAP) has published reports describing specific projects—perhaps the findings of 20 percent of the projects noted in this report are presented in more detail in reports published by NAP. And a few NAS and NRC press releases, newsletters of the Office of Central Europe and Eurasia, internal project documents, and web postings are in principle available for public scrutiny. Yet it is not easy for government officials or scholars, let alone the general public, to identify the specific documents that would be of interest. The RAS has very limited information readily available on the interacademy program.[3]

Only by directly participating in projects or interviewing the staff and participants who have been involved in projects is it possible for one to gain more than a superficial impression of the interacademy effort over many years, or even during the past decade. Unfortunately, because of budget limitations an earlier comprehensive review of past accomplishments and past difficulties was not possible. In the 1980s, limited summaries of program activities were presented to the U.S. Congress. The handful of more detailed reviews that have been conducted have usually been tied to renewal of contracts with federal agencies or applications for supplementary grants from private foundation, and thus they have been quite narrow in scope and buried in internal budget documents. This report, prepared with the support of private funds of the National Research Council, is intended to help fill the "evaluation" void common to many organizations absorbed in promoting action programs with near-term results.

Another aspect of this report that should be of interest is the inclusion in the text and the appendixes of extracts from several documents prepared by the RAS and the Russian government that relate directly to the objec-

---

[3] Usually the scientific secretary of the RAS includes in his annual report comments on interacademy activities. Summaries of his report are published each year in the journal of the RAS, *Vestnik*.

tives of interacademy cooperation. Dozens of Russian government decrees, strategy documents, and reports on the role of international cooperation in supporting Russian economic, social, and security goals, as well as references to some of the most important documents, also can be found in publications cited in this report.

Many American participants in interacademy programs, as well as in cooperative activities sponsored by other organizations, have been advocates of implanting in Russia elements of the U.S. model for the organization, management, and funding of science and technology. Certain concepts such as peer review, special tax considerations for nongovernmental organizations, and support for small, innovative businesses have begun to take root in Russia. However, there are many uncertainties about whether other elements of the U.S. model are appropriate for Russia, including, for example, establishment of a large number of research universities, widespread commercialization of university research, or support of only a limited number of research centers not aligned with universities. The purpose of this report is clearly not to join the debates on such important issues. They will undoubtedly continue to be the subject of discussion when leaders of Russian and U.S. science and technology meet under governmental, academy, or other auspices. And they will be discussed at meetings involving the leaders of France, Germany, Japan, and other countries that have their own models. In the end, Russia must develop Russian models, and international cooperation can only assist in identifying options that might be considered.

Finally, for the past several decades the Soviet and then the Russian systems of political governance and the frameworks for economic and social development have been undergoing dramatic changes, while domestic U.S. policies have been evolving much more slowly. Thus, in this report the background discussions of domestic developments that influence cooperation focus largely on changes in the Soviet Union and Russia and offer little information about the relatively stable situation in the United States. The foreign policies of both countries have undergone radical transformations, however, and these policies are treated in a more equitable manner.[4]

---

[4] For discussions of linkages between science and foreign policy, see, for example, Joyce (1982), Woodrow Wilson Center and Smithsonian Institution (1984), and Schweitzer (1988).

Acknowledgments

This report reflects the efforts of the thousands of American and Russian officials, scientists, and engineers who have been involved in scientific engagement between two of the world's largest countries. Some of these participants have been members of the academies of the two countries, others have been junior researchers who were destined to become members, but most have been less well-known specialists with strong technical capabilities essential for the efficient functioning of the science and technology infrastructures of the two countries. Dozens of administrative personnel also have provided essential support in organizing the details of activities of the participating specialists.

Deserving special recognition are the officers of the U.S. National Academy of Sciences, the Soviet Academy of Sciences, and the Russian Academy of Sciences, who, since the late 1950s, have consistently supported bilateral engagement and have included bilateral cooperation among the high priorities of their academies. In recent years, the presidents of the National Academy of Engineering and the Institute of Medicine have joined in this support. Also singled out for special appreciation are the leadership and staff of the National Science Foundation, which has provided financial support for scientific exchanges for almost a half-century. Moreover, dozens of other government agencies in the two countries have provided both financial and logistical support for interacademy programs, and private foundations have provided substantial financial contributions as well.

The following U.S. government departments and agencies, private organizations, and individuals (listed in order of magnitude of support) have

served as the principal funders for the National Academies in their cooperation with counterpart organizations in the Soviet Union and then Russia:
National Science Foundation, National Research Council, John D. and
Catherine T. MacArthur Foundation, Department of Defense, Agency for
International Development, Department of Energy, National Aeronautics
and Space Administration, Department of State, Carnegie Corporation of
New York, Nuclear Threat Initiative, The Rutter Foundation, Environmental Protection Agency, The Russell Family Foundation, National Institutes
of Health, Trust for Mutual Understanding, Rockefeller Family and Associ-
ates, and Office of Science and Technology Policy.

Having helped to guide the interacademy program for 16 years, I have
been in a unique position to witness the details of the formulation and
implementation of cooperative efforts, and I take full responsibility for the
content of this report. At the same time, many current and former members of the staff of the National Research Council and the RAS have provided substantial information that I have used here. Indeed, these staffers
have served as the backbone of the program and deserve special credit for
sustaining an ambitious and productive array of activities over many years.

I wish to thank the following individuals for their review of this re-
port: Kennette Benedict, John D. and Catherine T. MacArthur Founda-
tion; Loren Graham, Massachusetts Institute of Technology; Inta Morris,
U.S. Civilian Research and Development Foundation for the Indepen-
dent States of the Former Soviet Union; Victor Rabinowitch (retired),
John D. and Catherine T. MacArthur Foundation; Gerson Sher, U.S.
Civilian Research and Development Foundation for the Independent
States of the Former Soviet Union; and Alvin Trivelpiece (retired), Oak
Ridge National Laboratory.

Although the reviewers listed have provided many constructive comments and suggestions, they were not asked to endorse the report's conclusions or recommendations, nor did they see the final draft of the report
before its release. I am also indebted to consulting editor Sabra Bissette
Ledent for her assistance in transforming a hastily assembled and often
confusing document into a coherent and readable manuscript. As noted,
responsibility for the final content of this report rests entirely with the
author.

Contents

About the Author   xviii

1   U.S.-Soviet Scientific Cooperation in the Age of
    Confrontation   1
    Technology Transfer and National Security, 6
    International Security and Arms Control, 8
    Dissidents, Refuseniks, and the Exile of Andrey
       Sakharov, 9
    Reviewing the Early Record, 11

2   *Perestroika* and Expansion of Scientific Cooperation   15
    The Wider Program, 17
    Reflections on the Expansion of Cooperation, 26

3   Emergence of the New Russia: High Expectations,
       Harsh Realities, and the Path Ahead   30
    Weathering the Economic Crisis, 34
    Saving Russian Science, 35
    Reorienting the Interacademy Exchange Program, 38

4   National Security Issues and a Wider Agenda for
       Cooperation   41
    Protecting Nuclear Material, 43

Controlling Exports of Nuclear and Other Dangerous
    Materials, 46
The Wayward Weaponeers, 49
Redirecting Russian Biological Expertise from
    Military to Civilian Pursuits, 50
Counterterrorism on Center Stage, 55
Disposition of Spent Nuclear Fuel and High-Level
    Nuclear Waste, 58
Insights from Interacademy Consideration of
    Security Issues, 61

5   Supporting Innovation: From Basic Research to
    Payment for Sales   63
Individual Exchange Programs, 66
    Three Short-Lived Programs and One New Start, 66
    The Fifth Program: COBASE Continues Support
        of Basic Science, 68
Commercialization of Technology, 70
Improving Ethnic Relations in Russia, 74
Role of Russian Universities, 77
Global Environmental Problems, 77
Cooperation on Nonsecurity Issues: Lessons Learned, 79

6   Lessons Learned and the Future of the Interacademy
    Program   81
The View from Washington, 84
Insights from the Interacademy Program, 85
Learning from Reviews of Past Cooperative Activities, 85
    Cooperating on Important Topics Not Adequately
        Addressed in Moscow and Washington, 87
    Documenting Conclusions from Interacademy
        Projects, 88
    Engaging the Leaders of the Academies in
        Cooperative Activities, 89
    Encouraging Russian "Buy-in" for Concepts
        Developed Abroad, 89

Emphasizing the Sustainability of Short-Term
Projects, 91
Adopting Modest Goals for Interacademy Projects, 91
Interacademy Cooperation in the Years Ahead, 92

Epilogue   96

Appendixes

A   Highlights of Early U.S.-Soviet Scientific Relations
(1725–1957), 101

B   Agreement on the Exchange of Scientists between the
National Academy of Sciences of the USA and the
Academy of Sciences of the USSR (1959), 104

C   Agreement on Cooperation in Science, Engineering,
and Health between the U.S. National Academies and
the Russian Academy of Sciences (2002), 114

D   Agreement for Scientific Cooperation between the
Institute of Medicine of the USA and the Academy
of Medical Sciences of the USSR (1988), 117

E   Joint Statement by the Presidents of the U.S. National
Academies and the Russian Academy of Sciences [on
Preventing the Proliferation of Nuclear Weapons and
Nuclear Material], February 2, 2002, 122

F   Annex 2 to the Agreement on Cooperation in Science,
Engineering, and Medicine between the Russian Academy
of Sciences and the U.S. National Academies (2002), 125

G   Joint Statement by the Presidents of the U.S. National
Academies and the Russian Academy of Sciences
[on the Development of Knowledge-Based Economies],
February 2, 2002, 127

H   Cooperation Between U.S. and Russian Academies Encourages Russian Investments in Innovative Research, 129

I   Innovation in the Russian Federation (2001), 131

J   Personnel Trends in the Russian Academy of Sciences, 133

K   Innovation Projects of National Significance, 135

L   The Threats to Russia (View of the Ministry for
     Emergency Situations), 137

References   139

About the Author

Glenn E. Schweitzer has served as director of the Office for Central Europe and Eurasia (previously named the Office for Soviet and East European Affairs) of the National Research Council since 1985. From 1992 to 1994, he was on a leave of absence to serve as chairman of the Intergovernmental Preparatory Committee (PrepCom) for the International Science and Technology Center (ISTC) in Moscow and then as the first executive director of ISTC, which was established by the governments of the United States, European Union, Japan, and Russian Federation. Since 1989 he has written six books on U.S.–Russian scientific cooperation, the proliferation of weapons of mass destruction, and high-impact terrorism.

1

U.S.–Soviet Scientific Cooperation in the Age of Confrontation

*Soviet society is no longer insulated from the influence and attraction
of the outside world or impervious to the need for external contacts.*
Henry Kissinger, 1976

In 1955, with East-West political relations temporarily on the mend, the heads of state and then the foreign ministers of the Soviet Union, France, England, and the United States met in Geneva, where they discussed a program for increased contacts with the Soviet Union. The Soviets rejected the Western emphasis on multilateral approaches, but suggested that some of the proposals might be transformed into bilateral programs, and particularly cultural exchanges of individuals and groups. To this end, and after lengthy negotiations, the U.S. and Soviet governments signed an agreement in January 1958 that provided for a range of bilateral activities, and particularly reciprocal visits in the fields of education, culture, and information. The agreement included, at the Soviets' initiative, a provision for an interacademy exchange program to be worked out by the academies in the two countries. (The intergovernmental agreement was entitled "Agreement between the United States of America and the Union of Soviet Socialist Republics on Exchanges in the Cultural, Technical, and Educational Fields.")

Then in 1959, the U.S. National Academy of Sciences (NAS) and the Academy of Sciences of the USSR (ASUSSR) agreed to establish a formal program of scientific cooperation that would emphasize exchange visits by leading scientists. This agreement dramatically expanded the scope of the earlier scientific contacts that are briefly touched on in Appendix A. (Appendix B sets forth the text of the first interacademy agreement that was signed in 1959. For contrast, Appendix C provides the more diverse text for the interacademy agreement signed in 2003.) From 1961 to 1979, 10 additional two-year agreements were signed to continue the program of scientific cooperation until 1981. In that year, as discussed later in this chapter, adjustments were made in the program because of the internal exile in Russia of nuclear physicist and dissident Andrey Sahkarov, but the program continued nevertheless uninterrupted.[1]

Soon, the Bolshoi and Kirov ballet companies became familiar attractions in the United States. Leading American musicians, writers, and sports figures began touring the Soviet Union. American students enrolled at Moscow State University, while Soviet professors gave lectures on U.S. campuses. And portable exhibitions portraying life in each of the countries were erected in some of their distant cities.

Meanwhile, less ambitious technology-oriented exchanges were a priority of the Soviet government, but the U.S. government frequently withheld approval of these exchanges as bargaining chips to gain Soviet acquiescence to cultural and informational activities, which were the U.S. priorities. At the same time, the U.S. government hoped that the exposure of Soviet scientists and engineers, along with specialists from other fields, to U.S. achievements would contribute to the slow evolution of the Soviet Union in the direction of American society.

Short-term exchange visits of up to several weeks by individual scientists characterized the earliest years of interacademy cooperation. In general, the participants from both sides were highly qualified researchers. In the late 1960s, longer-term visits of up to one year became commonplace. The annual level of exchanges reached 167 person-months in each direction in the mid-1970s, but then declined to 50 person-months in the early 1980s because of budget cuts at the National Science Foundation, the financial sponsor.[2] Also in the 1960s, the two academies began to organize bilateral workshops on frontier topics in mathematics, physics, earth sciences, life sciences, and other disciplines. These workshops were highly visible events, and they served as signals to the scientific communi-

---

[1]For a discussion of the early days of U.S.-Soviet cooperation see Byrnes (1976: 76) and NAS (1977). A more recent review of exchange programs is presented in Richmond (2003).

[2] For a detailed discussion of the early interacademy exchanges, see Schweitzer (1992).

ties in the two countries that bilateral cooperation between political adversaries was acceptable. By 1981, 23 interacademy scientific workshops had been held.

Many American university-based scientists had hoped that the interacademy channel would be a nongovernmental channel relatively free of government interference. However, the governments were and will remain important participants in interacademy activities: academy institutes in Russia are government institutions; the NAS and many of the participating American scientists receive funding from the U.S. government departments and agencies that help support exchanges; and both governments monitor, and if necessary control, exchanges through the visa process. Constraints on the academy-to-academy channel persist today, but the political and administrative distances between government and academy activities in both countries in this arena are far greater now than they were two and three decades ago.

A specific example of the coupling of government and academy interests occurred in the early 1970s when Secretary of State Henry Kissinger successfully promoted a decade of expanded bilateral intergovernmental scientific and technological cooperation as one of the centerpieces of U.S. efforts to improve relations between the two countries. This cooperation was brought to life in 11 formal intergovernmental agreements in science and technology. For a few years, these agreements had the desired political effect of translating the concept of détente into highly visible activities (Ailes and Pardee, 1984; Schweitzer, 1989: 140–141). The governments selected the NAS and ASUSSR to lead the physics program, and a series of meetings and consultations involving leading physicists from both countries ensued over a period of more than 10 years. A good example of an important government initiative was the evolution of the intergovernmental Agreement on Peaceful Uses of Atomic Energy and its annexes. This agreement resulted from discussions between U.S. president Richard Nixon and Soviet president Leonid Brezhnev, and was then embedded in the Kissinger initiative. It led to hundreds of exchanges of scientific importance. Meanwhile, the core interacademy program of individual exchanges remained independent of these larger initiatives and continued. Indeed, the interest among scientists in the two academies expanded to additional fields, including science policy, the social sciences, and engineering.

Over the years, adjustments in the character and scope of interacademy cooperation have been driven by a variety of factors. They have included:

- early U.S. efforts to foster greater reliance on direct scientific contacts endorsed by the academies rather than on contacts brokered by the academies
- Soviet pressures to shift from an initial preoccupation with research in fundamental science to greater emphasis on industrial activities
- the emergence of related exchange programs, which resulted in redundant channels for cooperation
- concerns about exchange programs becoming mechanisms for a brain drain to the United States
- significant changes in intergovernmental political relationships that inevitably affected scientific activities.

Throughout the history of cooperation via academy and other channels, the linkage between the science and technology capabilities and the national security concerns of both countries has been strikingly evident. Because Russia is perceived as a declining international powerhouse, however, the interest of some funding organizations in the U.S.-Russia relationship has decreased correspondingly.

Prior to the 1990s most adjustments in the formal program structure were relatively minor and were easily made by the academies on an amicable basis, with the exception of the partial interruption of activities resulting from the Soviet treatment of scientist Andrey Sakharov. Thus, for several decades the interacademy relationship served as an important rudder of stability in the sometimes volatile relationship between the two countries. Also during this period when secrecy cloaked many scientific activities in the Soviet Union, the program provided an important channel through which American scientists could gain access to Soviet colleagues, facilities, and databases and through which Soviet scientists could make personal contacts with Americans whose names they had frequently seen on Western publications.

In December 1977 a review panel established by the NAS (often referred to as the Kaysen panel in recognition of its chair, Carl Kaysen of Princeton University) released its report on U.S.-Soviet interacademy exchanges and scientific relations (NAS, 1977). The panel set forth a detailed listing of the objectives of scientific exchanges, which are shown in Box 1-1. It then concluded that the 18-year-old interacademy program had been worthwhile and continued to be important, even in the era of much larger intergovernmental science and technology exchanges. After weighing the responses to surveys of more than 100 American participants in the

BOX 1-1
Objectives of U.S.-Soviet Scientific Exchanges

Building world science

- training young scientists
- communicating existing knowledge through lectures, meetings, symposia, and summer schools
- generating new knowledge through both short-term and long-term collaborative research
- developing a global strategy for scientific research in which each country would be able to optimize its research

Building U.S. science

- all of the points under "building world science" (above) with special emphasis on access to knowledge, materials, and techniques not otherwise available and to progress in fields in which the Soviet Union has superior or outstanding positions

Keeping abreast of Soviet science

- maintaining a continual awareness of Soviet scientific capabilities and Soviet resource allocations to science as a whole and within various fields

Fostering the international scientific community

- developing personal contacts
- encouraging scientists and scholars working under repressive or otherwise difficult conditions
- encouraging full Soviet participation in international science (e.g., conferences, international scientific bodies, adherence to international standard of scientific intercourse)
- gaining new cultural perspectives through both an increased knowledge of other cultures and the added insight such knowledge provides about one's own

Fostering the solution of global problems

- health, food, disarmament, energy, environment

(continued)

BOX 1-1 *(continued)*

Political objectives: using scientific and technological interchange as a way of reducing political tensions

- maintaining communications as such: links between scientists as leadership groups in the United States and the Soviet Union
- contributing to détente
- helping to maintain science as an essentially humane and liberal endeavor

Economic objectives

- promoting commercial exchange of technology and other trade related to scientific research

Source: Adapted from NAS (1977: 22–24).

interacademy program, the panel pointed out that while there had been substantial scientific benefits to the United States, even more valuable had been the less tangible benefits such as contributions to expanding the international scientific community and allowing U.S. scientists to keep abreast of new developments in Soviet science. The panel noted that American scientists had probably taught Soviet colleagues more than they had received in return, but it expected the balance to shift toward greater equality. As for U.S. government concerns about the transfer of sensitive technology, the panel did not find this issue particularly troublesome in view of the program's focus on basic research rather than on technological applications. Finally, the panel recognized the program's importance in providing a channel through which both the NAS and individual American scientists could express their concerns about human rights abuses in the Soviet Union.

TECHNOLOGY TRANSFER AND NATIONAL SECURITY

In response to increased interest, particularly by the Soviet government, in acquiring insights through exchanges revolving around the application of technologies to industrial problems, the interacademy program slowly expanded from basic research to encompass technological research and engineering innovations. Yet at the same time the U.S. government became more aggressive in thwarting repeated Soviet attempts to pilfer technological secrets that could bolster their military efforts. Increasingly, restrictions on U.S. visas for Soviet exchange visitors curtailed visits to specific laboratories and restricted the topics Americans could discuss with them while in the United States. Sometimes, U.S. visas were simply denied, despite appeals from the proposed American hosts who argued with little success that because the U.S. work was already published and the Soviet research was largely unknown, they would clearly gain from an exchange. Meanwhile, the efforts of the Federal Bureau of Investigation (FBI) to monitor the activities of Soviet visitors became more apparent to both NAS officials and the American hosts. At the same time, American visitors to the Soviet Union continued to report that Soviet security officials were interfering in the visitors' activities.[3]

During the early 1980s, many voices in Washington argued that scientific cooperation with the Soviet Union made little sense when that country was using every means at its disposal to gain a technological edge in all aspects of military science and technology. These advocates of scientific isolation of the Soviet Union, who were housed primarily in the Department of Defense, caused considerable confusion both within Congress and throughout the government about the appropriateness of scientific cooperation (Office of the Undersecretary of Defense for Policy, 1985).

Amid this controversy, the NAS established a special panel (often called the Corson panel in recognition of its chair, Dale Corson of Cornell University) to address the issue of scientific communication and national security, which was central to many government-sponsored and private exchanges. This panel issued its report in September 1982 with the following observations:

- There had been a substantial transfer of technology—much of it directly relevant to military systems—from the United States to the Soviet Union through diverse channels, but universities and open scientific communication had, in general, been the source of only an insignificant portion of the overall problem.
- Soviet efforts to acquire technology had increased in recent years, including efforts directed at universities and scientific research.

---

[3] Russian efforts to misuse the exchange program to obtain militarily sensitive technological data are discussed in Schweitzer (1989: 194).

- On occasion, the Soviets had used the interacademy exchange program for inappropriate purposes, giving information-gathering assignments to participants, who had in turn undertaken activities beyond the scope of their agreed fields of study.

- Imposition of controls could slow the rate of scientific advance and thus reduce the rate of U.S. technological innovation.

- Controls would impose economic costs for U.S. high-technology firms, affecting both their prices and their market shares in international commerce, and would limit university research and teaching in important areas of technology.

In summary, the panel concluded that, as a national policy, "security by accomplishment" had much to recommend it over a policy of "security by secrecy" (NRC, 1982).

The protection of sensitive technologies—the hardware, software, and technical data that provide the basis for designing and using new hardware—continues to be of central importance in organizing cooperative programs with Russia and many other countries. Often, the issue is joined when foreign specialists apply for visas to visit the United States. An up-to-date perspective on this topic is presented in Chapter 3.

INTERNATIONAL SECURITY AND ARMS CONTROL

As U.S.-Soviet relations deteriorated with the Soviet invasion of Afghanistan in 1979 and the Soviet downing of a Korean passenger airliner in 1983, leading members of the NAS and ASUSSR became increasingly concerned about the danger of nuclear war. The two academies designated groups of specialists with extensive personal experience in international security affairs to hold an exploratory meeting in June 1981 about the desirability of undertaking discussions on significant aspects of international security and arms control. The specialists agreed on the importance of the proposed initiative and developed the initial framework for a series of closed interacademy meetings dealing with substantive scientific and technical issues. The first such meeting was held in January 1982. Since then, 30 meetings have been held by the parallel Committees on International Security and Arms Control (CISAC) established by the two academies.

Clearly, these continuing dialogues on international security and arms control are an important aspect of the relationship between the Russian and U.S. academies. The high level of expertise of the participants, the steadfast commitments of the academies to obtaining financial support for the dialogues, and the interest of the governments in the views expressed and the conclusions reached attest to the significance of the effort.

Because these committees have considered a wide range of topics over the past 20 years, it simply is not possible to even highlight the dialogues in this short report, and so that task must be left to others. Nevertheless, it is important to keep in mind that these dialogues have been a stable element of the interacademy program and exemplify how the academies' unofficial channels of communication have complemented intergovernmental negotiations on important issues.

DISSIDENTS, REFUSENIKS, AND THE EXILE
OF ANDREY SAKHAROV

During the 1970s and 1980s, human rights became an important dimension of the U.S.-Soviet relationship (U.S. Congress, Commission on Security and Cooperation in Europe, 1988). Most of the concern within the American scientific community was centered on the plight of Soviet scientists who had been imprisoned for political reasons (the dissidents) and on these and others who had been denied exit visas (the refuseniks). The vast majority of these scientists were Jewish, but some from other groups, including German and Armenian, also were affected. Some American activists urged American scientists to refuse to cooperate with Soviet colleagues until the problems of the dissidents and refuseniks were resolved satisfactorily, and these activists clearly had an impact on the willingness of at least a few American scientists to receive Soviet visitors under the interacademy exchange program. Moreover, at times, leaders of the NAS appealed to their ASUSSR counterparts for information on the status of specific scientists believed to be imprisoned, and these formal appeals became a significant element of interacademy relations.

Of particular concern to the NAS was the Soviet government's treatment of scientist Andrey Sakharov, who had been elected a foreign associate of the NAS in 1972. He was repeatedly harassed and detained by the security services for his outspoken criticism of government policies and his role in organizing meetings of other human rights activists with similar views. Then, in 1979, the Soviet government exiled him to an apartment in the city of Gorky on the Volga River. In response, in February 1980 the Council of the NAS suspended for six months bilateral symposia, seminars, workshops, and new initiatives involving the ASUSSR or other Soviet organizations. This moratorium was later extended and remained the policy until 1985. The decision immediately affected activities planned in physics, science policy, and experimental psychology, and it placed a general damper on developing new workshop proposals that were in the formative stages (NRC, 1980: 1).

At the same time, however, the Council agreed that decisions about individual scientific visits were matters properly left to the consciences of the participating individuals, and the interacademy program of individual exchanges continued unabated. The new initiative in the field of international security and arms control, considered to be of the highest importance, also was exempted from the moratorium.

In the mid-1980s, the situation with the dissidents and refuseniks began to ease. Many were released from prison, and emigration to the United States, Israel, Germany, and other countries increased dramatically.

During this period, the Council of the NAS realized that its channels of communication with the ASUSSR had atrophied to the point that any hope of encouraging effective intervention by the ASUSSR in human rights cases was unrealistic. After considerable internal debate on the merits of a good channel of communication, even with the continued exile of Andrey Sakharov, the Council decided in 1985 to reestablish a broad program of bilateral scientific cooperation (NRC, 1985a: 1).

The ASUSSR gradually began to respond to appeals from the NAS for information about specific scientists of concern, but their responses were usually not encouraging. The leaders of the ASUSSR took the position that they would intervene only in cases involving scientists who had been academy employees, a group that represented less than 10 percent of the scientists on lists of dissidents and refuseniks prepared in Washington. Another frequent Soviet response was that many of the scientists being denied travel documents had access at an earlier time to state secrets and therefore could not be allowed to go abroad. This answer frustrated American colleagues because there was no Soviet policy on how long the scientists needed to wait before traveling after termination of such access. Meanwhile, ASUSSR officials privately commented to visiting Americans that they would like to see all the refuseniks leave because they were not contributing to Soviet science, and the Americans responded that the refuseniks wanted to leave precisely because they were denied opportunities to contribute to world science.

By 1988 the situation had changed significantly. Thousands of refuseniks were being given permission to leave each year. At a dramatic press conference in Moscow featuring the president of the ASUSSR and the president of the NAS, the Soviet president became the first Soviet official to acknowledge publicly that the Soviet record in human rights was not satisfactory and that many scientists had suffered from inappropriate treatment.[4] This act probably represented the most significant direct impact of the NAS position on human rights. At about the same time, Andrey Sakharov returned to Moscow from exile in Gorky, and he met with the visiting officials of the NAS.[5]

Human rights remain an indelible aspect of U.S.-Russian relations, and cases of mistreatment of Russian scientists remain on the agenda of the National Academy of Sciences, National Academy of Engineering (NAE), and Institute of Medicine (IOM). Fortunately, only a handful of egregious cases emerged during the 1990s. Russian security services will continue to have the upper hand in cases involving alleged espionage, but the RAS will nevertheless be a significant channel for expressions of Western concerns about human rights violations.

REVIEWING THE EARLY RECORD

During the 1970s and early 1980s, more than 25 evaluations of U.S.-Soviet cooperative efforts in science and technology activities were carried out by the U.S. government (individual departments and agencies, White House offices, Congress), NAS, and other organizations in the United States. The conclusions of these evaluations were generally consistent, and several important observations can be summarized as follows:

• Bilateral communications faced many obstacles, including lack of reciprocal access to specialists and facilities, concerns about human rights abuses, logistical problems, language barriers, and the inertia of bureaucrats in both countries who were not committed to cooperation.

• The high quality of Soviet research in certain fields—such as theoretical physics and mathematics—made cooperation in those fields extremely worthwhile for the United States as well as for the broader international scientific community.

---

[4] A photograph taken at the press conference, together with commentary on many developments related to human rights and other issues in Russia during the 1980s, are included in the memoirs of Guri Marchuk, the former president of the ASUSSR, who also served as vice premier for science and technology of the Soviet Union (see Marchuk, 1995).

[5] For a discussion of human rights issues and their impact on scientific cooperation during this era, see Schweitzer (1989: Chap. 8).

•   The most successful projects were highly focused, with specific objectives, and in fields in which there was a general parity of expertise.

•   Although there were many intangible cultural benefits from cooperation in science and technology, such benefits should not be the sole justification for cooperation.

•   The stable and open channels of communication played a very valuable role in efforts to stay abreast of Soviet achievements and to identify Soviet scientists who could contribute to overall international scientific efforts.

•   In general, the leakage of militarily sensitive technology, or knowhow, to the Soviet Union through bilateral scientific exchanges was minimal, largely because there had been little cooperation in areas of military significance. Security restrictions on scientific interchange were not only unwarranted in most areas, but also in some cases were detrimental to U.S. interests.

•   The erratic funding of bilateral activities, reflecting the ebb and flow of the overall relationship between the United States and the Soviet Union, had a negative impact on the effectiveness of cooperation, and consistent funding insulated from political vagaries was crucial to the future success of bilateral programs.[6]

Related to these evaluations was a continuing effort within the U.S. government to identify areas of particular interest for cooperative programs. An example of one attempt to identify such areas appears in Table 1-1, which compares the technical strengths of the two countries in selected areas of science.

As the adversarial nature of the U.S.-Soviet relationship began to change, the U.S. government redirected to other topics its handful of specialists who had been assigned the task of analyzing systematically civilian-oriented science and technology activities in the Soviet Union and then Russia. Soon, the U.S. government was relying heavily on academics who had participated in National Research Council (NRC) activities to attend workshops and meetings that provided authoritative insights into Russia's technical capabilities in order to fill an analytical void within the U.S. government. Workshops organized by the Department of State at Meridien House in Washington have been of particular importance in addressing issues such as

---

[6] This evaluation by the National Research Council staff was reported in NRC (1987d: 6–7).

TABLE 1-1   Relative Strength of Soviet Union in Specific Fields of Science Compared with the United States

Field	Comparable	Weaker
Mathematics	X→	
Atmospheric physics		X
Oceanology		
Theoretical	X	
Experimental		X
Materials science	X	
High-energy physics		
Theoretical	X	
Experimental		X→
Fluid dynamics	X	
Condensed matter physics		
Theoretical	X	
Experimental		X
Astrophysics		
Theoretical	X→	
Experimental		X→
Molecular biology		←X
Laser physics		
Theoretical	X	
Experimental	X	
Computer science		X

Note: An arrow indicates the probable direction of change in the future relative status where such an indication can be made with reasonable confidence. Some fields have not been divided into theoretical and experimental because of lack of data.

The Soviets were found not to be grossly stronger than the United States in any of these fields.

Source: OSTP (1985: 18).

the evolution of a knowledge-based economy in Russia (workshop held in August 2001).

In addition to providing scientific enrichment for both countries, U.S.-Russian scientific cooperation through many channels over several decades undoubtedly contributed to the eventual unraveling of the Soviet Union. The sharp contrast between the openness and rigorous peer review of research activities in the West and the inward-oriented research approaches in the Soviet Union made an impression on many Soviet officials and researchers, who began to question the compatibility of authoritarianism and scientific progress. Then in the 1980s, Soviet visitors to the West witnessed how personal computers were becoming standard equipment in offices and laboratories, while Russian schools could not even offer hands-on computer experiences for students. In the well-known Siberian science city of Akademgorodok, for example, the schools were able to find a few computers only because concerned parents employed at the Computer Center of the Russian Academy of Sciences diverted to the schools computers that would not be missed during inventories (NRC, 1988c: 3). In short, Soviet political and scientific leaders became painfully aware that their centralized planning system was not in tune with the more effective approaches to managing technologies that were fueling economic growth around the world.

2

Perestroika and Expansion of Scientific Cooperation

Facing the truth and publicly debating the nation's most acute
and vexing difficulties are supposed to be the strength
of democracy. Isn't it extraordinary that this has been happening . . .
in the land of Stalin and Ivan the Terrible?
Hedrick Smith, *The New Russians,* 1991

By the mid-1980s, Mikhail Gorbachev's efforts to reform the Soviet system of governance were well under way. *Glasnost* (openness) was becoming a reality as the traditional dearth of information about developments in the country was replaced by an overload of information of all shades of reliability. Soviet investigative reporters began exposing corruption, inefficiencies, and structural weaknesses throughout the Soviet state. Soviet army troops had returned from Afghanistan, and excessive military expenditures had increasingly become the target of criticism from within the government and in the press. Private restaurants, bakeries, and repair shops were springing up in cities throughout the country, as the government heralded the establishment of small private enterprises as an important mechanism for absorbing some of the underutilized workforce within the country (Kaiser, 1988/89).

Meanwhile, consumer goods were becoming scarce in every city and town. Even vodka production became a victim of the new thinking. But when the population rose up in protest over the long lines to buy a bottle of vodka and bootlegged vodka became increasingly popular, the government reversed its policy intended to improve the health of the population by limiting consumption of alcohol and authorized increased production.

Even the best Soviet industries lost their competitive edge as imports of modern technologies rose while the value of the ruble tumbled. To further exacerbate the situation, the nuclear reactor tragedy at Chernobyl, the earthquake devastation of poorly constructed buildings in Armenia, several gas pipeline explosions, and increasing shortfalls in agricultural production shook the confidence of the population in Soviet technology, which had for decades been a symbol of the strength of the Soviet system. The repairs needed to remedy these and other catastrophic failures of technology, stemming largely from shoddy Soviet practices, drained scarce resources (Garrett, 1988; Schweitzer, 1989; Graham, 1993).

Gorbachev repeatedly called on intellectuals, and particularly scientists and economists from the ASUSSR, to help find practical solutions to these and other problems impeding economic growth. The new government advisers urged adoption of Western management approaches. And they recommended the immediate purchase of tens of thousands of computers in an effort to energize the entire society. They also called for reorientation of some of the technologies that had supported the large military effort to the challenging task of upgrading industrial production practices in the civilian sector.

In the foreign policy arena, Soviet academics became significant participants in government entourages at disarmament talks in Geneva and at other important intergovernmental gatherings. They did speak out against the U.S. Strategic Defense Initiative (Star Wars), but they also offered all nations practical approaches to reducing the nuclear threat through arms reduction and through steps to prevent nuclear proliferation (Sagdeev, 1994).

A truly heavy burden to provide the conceptual basis for a revitalized nation had been placed on the doorsteps of leading Soviet scientists. Some of these scientists were also the Soviet interlocutors for the NAS-ASUSSR interacademy program.

Many of Gorbachev's advisers urged closer cooperation with U.S. institutions, both to help reduce international security tensions and to take advantage of Western experience in competing in international technology markets. In 1987 and again in 1990 several of his academic advisers accompanied him to summit meetings in Washington, and special sessions were arranged for them at the NAS. The agendas for these meetings, which were very rich, included highly informative discussions of economic reform, the legal framework for *perestroika*, international cooperation in space research, the use of computers in education, and other topics (NRC, 1988e: 4, 1990a: 1).

Against the changing political and economic panorama in Moscow, a general warming of bilateral relations between the two governments, and the need for better channels to address human rights issues, the leadership of the NAS decided to reestablish a broad program of cooperation with the ASUSSR in 1985 (NRC, 1985a: 1). The National Science Foundation (NSF) was prepared to continue to provide substantial funding for exchanges of individual scientists to and from the Soviet Union—on the order of $400,000 annually to support 50 person-months of exchanges in each direction, staff support costs, and related activities. In addition, in 1984 the NAS had received a 10-year grant of $3 million from the John D. and Catherine T. MacArthur Foundation to support activities with colleagues in the Soviet Union and China, and these funds enabled the NAS to consider a variety of new program approaches (NRC, 1983–1984a: 3). The funds were transferred to the NAS at the outset of the 10-year period, and their investment earnings eventually increased the available funding to over $5 million, with the bulk of the funds devoted to the Soviet engagement program.

In Moscow, there was an eager response to the renewed interest in Washington in interacademy cooperation, but only as long as the focus was on scientific activities and not on human rights and other contentious issues. A brief clause in the new interacademy agreement in 1986 that was acceptable to both sides was clearly directed to concerns over human rights ("the environment affecting cooperation"), and within a few months many new cooperative activities were under way.

THE WIDER PROGRAM

Not surprisingly, one of the new areas for interacademy consultations was economic reform.[1] Meetings of specialists were held in the Soviet Union and the United States to consider the changes brought about by *perestroika*. The topics for the first meeting in Moscow in 1987 included patterns and trends in economic structure and aggregate productivity, the economic aspects of technology, innovation and the diffusion of technology, management approaches, and international economics. Despite the timeliness and the importance of the items on the agenda, the presentations by Soviet econo-

---

[1]For a detailed discussion of economic reform efforts, see U.S. Congress, Joint Economic Committee (1987).

mists wedded to central planning provided few new insights into the problems confronting a country presumably interested in transition to a market economy. Also, considerable confusion surrounded the selection of the appropriate American and Soviet participants for the dialogue. In particular, Soviet efforts to involve their leading engineers in meaningful discussions of macroeconomic issues were not successful (NRC, 1989a: 11).

At the second meeting in the United States, there was a better match of economic expertise between the Americans and Soviets, and the agenda was expanded considerably. Although the discussions continued to be superficial, they provided an impressive overview of the many topics high on the Soviet list of economic priorities. These topics included conversion of defense industries, the new law on private property, the speed of the process of *perestroika*, personal savings and investment, the need for a convertible ruble, Soviet interests in stocks and bonds, joint ventures, and the threat of labor strikes. A few Soviet economists were beginning to think like Westerners, and the American participants began to understand the large gap between Western and Soviet conceptions of economic policy in a free market economy (NRC, 1989–1990a: 16).

The academies also sponsored two-week workshops for young economists from the two countries in the United States and in the Soviet Union. The promising young scholars who participated addressed a variety of interesting topics, including enterprise reform, environmental economics, the innovation process, internal currency markets, and price reform. Several of the participants kept in touch after the workshops, and they soon gained recognition from their American and Soviet peers within and outside government for their insightful views of the Soviet economic transition (NRC, 1990f: 5).

Then in 1990, several U.S. government agencies asked the NRC to assemble a group of experts to help estimate the size of the Soviet economy, an issue immersed in controversy in Washington since 1980 and even earlier. The group concentrated on the military-industrial sector, consumption and service activities, and the underground economy. Participants quickly concluded that Soviet statistics were unreliable. They simply could not venture a conclusion to meet their charge, and they urged the Soviet Union to adopt an internationally acceptable accounting system as an indispensable component of its transformation (Alexeev and Walker, 1991).

Earlier, in 1986, the two academies had initiated a series of interacademy meetings on energy efficiency and conservation, another area of immediate policy concern. This topic was and remains of crucial importance to both countries, and the specialists identified a large number of topics that could be effectively addressed by American and Soviet specialists (NRC, 1986a, 1989b: 9, 1989d: 5). Box 2-1 identifies some of the energy efficiency and conservation topics suggested for interacademy cooperation. A particularly interesting aspect for the American participants in this initial collaboration was a 10-day tour of key energy facilities in Siberia—hydro projects, energy transmission facilities, mining centers, and research institutes (NRC, 1989–1990b: 15).

Ecology projects also became important components of the interacademy program. Global ecology was a topic of interest to both scientists and engineers, and, within that context, biodiversity and the impending extinction of thousands of plant and animal species were popular themes for exchange visits by groups of specialists. In meetings attended by representatives of regulatory authorities, the academies addressed environmental monitoring, the health effects of pollutants, ecological resiliency, and global and regional-scale studies of environmental change. Meanwhile, the overarching concept of "environmental security" was adopted in both countries, and the academies explored the far-reaching effects of neglect of air and water pollution, indiscriminate cutting of forest areas, and degradation of soil.

Social scientists took advantage of the expansion of opportunities for new areas for cooperation. Initially, the academies sponsored a workshop on social science research and the prevention of nuclear war, paying particular attention to various concepts of interdependency and modeling of interdependency. Global ecological problems also were on the agenda. Another workshop topic was research priorities in improving understanding of the challenges of northern regions where living conditions are very difficult. The social scientists urged expansion of joint projects and resolution of the logistical issues that impeded cooperative efforts to help develop the behavioral sciences in the Soviet Union (NRC, 1987c: 7, 1990d, 1990e).

Two tragedies in the Soviet Union—the Chernobyl accident in 1986 and the earthquake devastation in Armenia in 1988—triggered quick responses by the NAS. Over a period of several years, the NAS sponsored visits by American specialists to Chernobyl and to bilateral and international meetings to assess the extent of the contamination problems and the effects both of the accident itself and of the contamination on human health and on animals and crops. As for the earthquake, the NAS, in cooperation with the U.S. Geological Survey, sent a team of earth scientists to Yerevan to review the damage and the likelihood of aftershocks. The general objective

BOX 2-1 Energy Efficiency and Conservation Topics Suggested for Interacademy Cooperation, 1986

Specialists of the National Academy of Sciences and the Soviet Academy of Sciences developed this list of priorities for interacademy cooperation at an interacademy workshop in Yalta in 1986. The two selection criteria were likely impact on energy consumption and comparable technical strengths of the two countries.

Buildings
- thermal characterization of building components and systems
- fenestration
- field evaluation of energy use in buildings

Community systems
- advanced concepts in district heating

Transportation
- electric urban mass transport
- heavy-duty vehicle diesel engines

Industrial processes
- welding technologies
- melting and hot working of materials
- industrial coprocessing of energy and materials

Power generation and distribution
- advanced systems for cogeneration, especially with gas turbines
- improvement of energy equipment—for example, generators, turbines, boilers, and transformers
- power conditioning
- high-voltage transmission
- superconductivity applied to power generation and transmission

Energy demand analysis and modeling
- integration of technology, economic, and environmental concerns to improve the modeling of energy systems and the forecasting of energy demands

Basic science to support energy conservation
- heat and mass transfer
- tribology
- combustion research

Source: Adapted from NRC (1986b: 7–8).

of the visit was to improve understanding of earthquakes and their effects on engineering designs and construction of structures. The activities stemming from this objective strengthened the scientific ties between specialists, which contributed to a high level of cooperation on earthquake prediction and seismic engineering in the years ahead (NRC, 1989c: 6).

Concerned about other harmful incidents that could be produced by nuclear activities, the two academies began organizing workshops on nuclear topics, and such efforts continue to this day. Risk assessment and reactor safety received greater attention worldwide after the Chernobyl accident, and they became useful focal points for interacademy discussions. The topic of radioactive waste management was becoming a contentious issue in both countries, and two workshops were directed to this topic. Also, in view of the fact that the Armenian earthquake occurred in the region of a nuclear power reactor, the academies gave considerable attention to the impacts of external events on nuclear reactor safety (NRC, 1988g: 17, 1989–1990c: 13, 1990c: 6).

But these nuclear-related activities did not plow much new ground because the same topics were being addressed on a much broader scale within cooperative programs sponsored by the U.S. Department of Energy and the Soviet Ministry of Atomic Energy (Minatom). Yet the interacademy deliberations were important in helping to establish the ASUSSR, and eventually the RAS, as an important participant in nuclear debates in Russia. A new Nuclear Safety Institute was established within the framework of the ASUSSR shortly after the Chernobyl tragedy. Located outside the Minatom complex, the institute slowly gained credibility as a "watchdog" institution.

The topic of science education also was on the list of priorities of the two academies. The lack of computer literacy in Russia was of special concern. In the United States, the inconsistent performance of high schools in science and mathematics was attracting attention. However, despite many discussions of these topics at interacademy meetings, a significant interacademy program did not materialize. The academies were not able to define a meaningful project because the discussions seemed to always slip back to those of how to obtain more funding for computers in the Soviet Union.

Even though much of the attention of the two academies had been redirected to applications of science, the importance of basic research was not forgotten. In 1986 the two academies identified eight scientific topics for bilateral workshops that were held during the late 1980s. As indicated in Box 2-2, most of the workshops were considered successes by the participants. The financial sponsors also were generally pleased.

BOX 2-2 Evaluation of Bilateral Scientific Workshops held from 1987 to 1989

This evaluation was based on reports by the NAS workshop chairs and staff evaluations. The significant criteria for evaluation of the workshops included the relative importance of the topic chosen for study, the degree of U.S. and Soviet strengths in the field, the composition and balance of the U.S. and Soviet scientific teams, and follow-on activities resulting from the workshop.

Workshop	Summary of Evaluation
Condensed Matter Theory *December 1987,* *Santa Barbara*	*** Outstanding. Became a successful facilitative project.
Lasers in Linear and Nonlinear Photochemistry *March 1988, Santa Barbara*	*** Highest level in laser research.
New Approaches to the Creation of Vaccines April 1988, Moscow	- - Opening of dialogue, but U.S. and Soviet participants had very different orientations.
Nonlinear Processes in Dense Plasmas *May 1988, Santa Fe*	** Major success; collaboration began.
Earthquake Prediction *October 1988, Moscow*	*** Established connections between seismologists and nonlinear experts; collaboration began.
Planetary Sciences *January 1989, Moscow*	** Very good workshop, but disappointing site visits.
High-Energy Astrophysics *June 1989, Tbilisi*	* Delegations not well matched, but workshop started collaboration.
Plant Molecular Biology Applied to Agriculture *October 1989, Washington,* *D.C.*	** Workshop started good scientist-to-scientist contacts.

(continued)

BOX 2-2 *(continued)*

Structure of Eukaryotic Genome *October 1989, Tbilisi*	** Very good workshop, but not enough young Soviet scientists.
String Theory *October/November 1989, Princeton*	*** Very high-level workshop; good exchange on intersections of mathematics and physics.

Note: Reports were published for two of the workshops: planetary sciences (Donahue, 1991) and high-energy physics (Lewin et al., 1991).
Source: NRC (1990b: 24).

One unsuccessful workshop was on the topic of virology. The American participants were disappointed with the level of scientific achievements presented by Soviet specialists during the workshop and subsequent visits to several research institutes. A few years later, they learned that a large sector of the Soviet biological research establishment had been excluded from the workshop. A huge complex of institutes and production facilities that had been established to support the Soviet defense effort—the Biopreparat complex—was simply under wraps and was little known to Western visitors even though the basic research capabilities of some of the Biopreparat institutes were precisely the types of capabilities under discussion at the workshop.

Indeed, access to former Soviet defense-related facilities has remained a frustration for American advocates of scientific exchanges. Similar access problems do exist in the United States, but in Russia the military legacy of secrecy lives much longer, beyond the termination of sensitive activities. The people who work in such facilities are so accustomed to secrecy that they have little incentive to challenge security procedures—no matter how much out of date.

A new aspect of interacademy cooperation during the 1980s was the facilitation of bilateral projects developed and carried out independently by individual scientists or institutions in the two countries. The role of the academies was simply to endorse the activities, assist with the acquisition of visas, and sometimes support efforts to obtain funds for the activities. Table 2-1 identifies facilitative projects selected for special attention in 1990. In general, these "bottom-up" activities were considered quite useful from the

TABLE 2-1 Facilitative Projects Endorsed by NAS-ASUSSR for 1990

Topic	NAS Lead	ASUSSR Lead	Status
Lasers in photochemistry	University of California at Los Angeles	Institute of Spectroscopy	Workshop held in Soviet Union in June 1990.
Algebraic geometry	University of Chicago	Steklov Institute of Mathematics	June 1989 symposium proceedings published in 1990. Individual visits made by Soviets to U.S. in 1990. Second workshop held in 1991.
Laser optics of condensed matter	University of California at Irvine	Institute of General Physics	Workshop held in U.S. in 1990.
Geological impacts and mass extinctions	Lawrence Berkeley Laboratory	Institute of Spectroscopy	Ongoing joint research conducted to determine possible extraterrestrial origin of geological sediments. Joint publication in 1991.
Solar neutrinos	University of Pennsylvania, Los Alamos National Laboratory	Institute for Nuclear Research	Soviet teams visit University of Pennsylvania and Homestake Gold Mine in 1990. Developed new iodine detector experiments.
Condensed matter theory	University of Illinois	Landau Institute of Physics	Two Soviet postdoctorates at University of Illinois and one at University of Florida.
Applications of nonlinear dynamics to problems of lithosphere deformation	Cornell University	Schmidt Institute of Physics of the Earth	Bilateral consortium established to oversee cooperative research.
Algebraic groups and related number theory	Princeton University, Yale University	Institute of Mathematics in Minsk	Framework developed for follow-on workshop.

Source: NRC (1990b: 25).

viewpoint of both the participants and the financial sponsors. But within several years, the academies were no longer needed to help facilitate such activities. The only impediment to the expansion of such projects was the lack of funds.

In 1988, in response to the personal interests of several members of the Institute of Medicine, the IOM and the Soviet Academy of Medical Sciences entered into a formal agreement for cooperation in four areas: alcoholism and chemical dependency, virology and host response to the human immunodeficiency virus (HIV), the application of molecular biology to the eradication of poliomyelitis, and the health effects of environmental radiation. This IOM initiative was about two years ahead of a closely related initiative of the U.S. Department of Health, Education and Welfare, which signed an array of agreements with the Soviet Ministry of Health to address the same issues as well as a few others. This onetime series of IOM exchange visits clearly helped to jump-start cooperation by focusing attention on both sides of the ocean on common problems of growing importance worldwide (NRC, 1988b: 2).[2]

The late 1980s saw an upsurge of interest within the two governments and in the U.S. private sector in expanding intergovernmental bilateral cooperation in science and technology. There were fewer visa and access problems than in previous years, and funding for cooperative activities had become more plentiful, at least on the U.S. side. However, the NAS soon lost an important advantage in competing for funding—its unique access to important pockets of Soviet science. In searching for a new unique role, the interacademy program quickly stretched beyond the traditional exchanges of individual scientists, as noted earlier.

Yet exchanges of individual scientists remained a core activity of the two academies, producing many notable achievements in addressing specific scientific challenges. Some examples follow:

- An American-Russian team obtained evidence that Siberia was on the equator during the Cambrian period (NRC, 1982–1983: 4).
- An American mathematician working with colleagues in Leningrad proved a solution to the Bieberbach Conjecture (NRC, 1985b: 3).
- An American-Russian team developed new comparative geoformation information on the Baykal and Rio Grande rift systems (NRC, 1988a: 10).

---

[2]The text of the agreement appears in Appendix D.

• An American-Russian team working in both countries provided new insights on alpine florae in North America from studies of their antecedents in Russia (NRC, 1988f: 13).

• An American geologist joined a Russian team that discovered a Kornerupien locality in Siberia with rock specimens comparable to the best in the world (NRC, 1987a: 8).

Impressive findings in cell biology, ice physics, and many other disciplines also attest to the importance of scientific engagements sponsored by the academies. And not to be overlooked are the contributions of scientists to political rapprochement. In the words of the U.S. ambassador to Moscow in 1983 after a presentation at his residence by a planetary scientist who was an NAS exchange visitor, "His trip helped NASA's reputation gleam more brightly here" (NRC, 1983–1984b: 2).

Meanwhile, in Moscow the ASUSSR was reaching out to many U.S. organizations. Table 2-2 identifies the most ambitious efforts of the ASUSSR in the 1980s. Still, both academies continued to give their special relationship high priority, even though cooperative projects sponsored by other organizations but involving the academies were often far larger. During the late 1980s, each of the academy presidents made two transatlantic visits at the invitation of his counterpart to discuss this special relationship.

A particularly important development for international cooperation in science and technology was the establishment in 1986 of a special science and technology committee by the U.S.-USSR Trade and Economic Council. Several of the projects promoted by the committee were directly related to the interests of the NAS and the ASUSSR (computers in education, energy conservation). Members of the committee from several U.S. companies such as Corning, Ralston Purina, and Monsanto that were particularly interested in the research activities of the ASUSSR and the Soviet Academy of Medical Sciences participated in the NAS-ASUSSR workshops.

REFLECTIONS ON THE EXPANSION OF COOPERATION

The proliferation of cooperative activities through the interacademy channel and through many other channels clearly indicated that the days of central planning of bilateral cooperation had come to an end. Government organizations and individual scientists themselves in both countries were determining whether their involvement in cooperative programs would be

TABLE 2-2 U.S.-Soviet Cooperative Programs Administered by the
Soviet Academy of Sciences during the 1980s

Field	U.S. Institution	Lead Institution within Soviet Academy
Portions of intergovernmental agreements		
Fundamental properties of matter	Department of Energy	Depends on project
Environmental sciences	Environmental Protection Agency, National Oceanic and Atmospheric Administration, Department of the Interior	Depends on project
Independent agreements		
Physics, chemistry, and materials sciences	National Bureau of Standards	Depends on project
Social sciences	American Council of Learned Societies/International Research and Exchanges Board	Depends on project
Organic chemistry	University of Minnesota	Zelinsky Institute of Organic Chemistry
Planetary geochemistry	Brown University	Vernadsky Institute of Geochemistry and Analytical Chemistry
Mechanical engineering	American Society of Mechanical Engineers	Blagonravov Institute of Machine Science
Decision theory	University of California at Los Angeles	Computer Center (Moscow)
Verification of nuclear testing	National Resources Defense Council	Schmidt Institute of Physics of the Earth
Physics	University of Texas	Lebedev Institute of Physics

Source: NRC (1988d: 7).

of benefit from their vantage points. Although the U.S. Department of
State and the Soviet Ministry of Foreign Affairs obviously remained significant organizations in the approval of activities, they had in large measure,
but not entirely, become the scorekeepers rather than the controllers of
cooperation.

Most of the joint activities during the late 1980s were based on targets of opportunity rather than a deliberative process to carefully weigh alternative project opportunities. The focus of the interacademy program shifted from exchanges in the basic sciences to workshops, consultations, and other forms of cooperation in areas of immediate security, economic, and social concern within the Soviet Union.

At the request of the U.S. Congress, in 1986 the president of the NAS had laid out general principles for cooperation with the Soviet Academy of Sciences in the new political environment. They included:

• an emphasis on projects in fields in which both countries were world leaders
• a new focus on scientific problems of global dimensions, particularly problems in the environmental and atmospheric sciences
• concentration of activities in nonsensitive areas
• greater access to unique data banks and scientifically important geographic areas
• consistency of policies of counterpart organizations with the provisions of the Helsinki Accords (particularly human rights provisions), with annual discussions of the means to foster cooperation within this framework
• more active participation in cooperative endeavors by leading scientists, and particularly academy members, than in the past
• more exchange visits based on invitations from foreign colleagues, in contrast to the earlier system of nomination of exchangees by the respective academies (NRC, 1987d: 6).

That same year, the two academies developed new initiatives reflecting these principles. As the political and economic transitions began within the Soviet Union, policy-oriented activities moved to center stage, with cooperation in the basic sciences receiving less attention. From 1980 to 1991 the academies sponsored 20 workshops, evenly divided between those on scientific research topics and those on policy-oriented topics.

The attraction of travel to the Soviet Union waned during this period. Procedures in Moscow covering the timely issuance of visas and Soviet support for the organization of itineraries became uncertain. The quality of hotel accommodations also deteriorated. Indeed, the general level of amenities in the Soviet Union declined throughout the travel sector.

At the same time, the U.S. government was promoting exchanges of all types on the grounds that Soviet exposure to Americans would contribute to positive changes in the approach to governance in the country. Advocates of scientific cooperation argued that the objectivity and openness of scientific research were characteristics that would contribute to such positive changes. But this political motivation for engagement never supplanted scientific benefits as the primary rationale for contacts (U.S. House of Representatives, 1986).

In summary, during the late 1980s the academies were major players in pioneering new areas for cooperation and in testing new mechanisms for its implementation. Also, the academies included in their activities many specialists who had not previously been interested in cooperative projects. Some of these participants went on to play important long-term roles in fostering U.S.-Soviet and then U.S.-Russian relations. The payoffs from and limitations of scientific cooperation through academy and other channels had become clearer (Ailes and Pardee, 1984).

3

Emergence of the New Russia:
High Expectations,
Harsh Realities, and the Path Ahead

We are losing an entire generation of young scientists.
We do not support them, and they think only about money.
Russian Minister of Science and Technology, 1998

The year 1991 was a tumultuous one in Moscow and throughout the former Soviet Union as the country, facing irreversible political cleavages and difficult financial problems, splintered into 15 independent states. The budgetary resources and the financial obligations of remnants of the former Soviet state immediately became the subject of disputes at the intergovernmental level, among commercial and noncommercial organizations, and between individuals who thought they had first rights to assets that suddenly belonged to everyone or to no one. Not surprisingly, in a country suddenly bereft of affordable consumer goods, thousands of newly installed government officials and holdover enterprise managers in all regions of Russia directed their energies to acquiring personal control over resources of questionable ownership, thereby buttressing their own well-being.

Amid the scramble for rapidly evaporating assets, continued financial support for researchers and for other scientists was not a priority within the new governments or among the general populations of the 15 states. Most Soviet scientists resided in the Russian republic, where they faced the challenge of competing with far more financially savvy elements of society for their share of the remnants of the Soviet state. The industrial base was eroding rapidly, and many Russian organizations found it difficult even to meet payroll obligations. In particular, the leaders of hundreds of enterprises that had traditionally supported substantial research and development activities, both within the enterprises and through outsourcing with other organizations, quickly lost interest in financing research projects that did not have an immediate payoff. The average salaries of Russian researchers sank to the equivalent of $25 a month, and growing numbers found gardening at their dachas more challenging and profitable than toiling in their laboratories without access to electricity, supplies, or scientific publications.

Historically, each of the 15 Soviet republics, with the exception of the Russian republic, had had its own academy of sciences which received "scientific guidance" from the ASUSSR—guidance usually extended to control of budgets and personnel appointments. Meanwhile, the ASUSSR had served as the de facto academy for the Russian republic as well as the parent academy for the entire Soviet Union. Thus, for decades the ASUSSR was able to concentrate large efforts at geographically dispersed research centers on centrally determined priority programs. Many scientists in the outlying republics considered the entire academy structure to be simply a mechanism to ensure Russian control over scientific activities throughout the Soviet Union, while giving the appearance that Moscow recognized the importance of local autonomy in addressing problems of special interest to the republics.

Within this context of long-standing centrally controlled research, rampant financial chaos, and regional suspicions over the motivation of Moscow's science administrators, the future role and structure of all elements of the ASUSSR became a highly politicized issue. In Russia, most academicians rallied together to preserve the professional and financial benefits of the monthly honorarium that had accompanied their membership in the ASUSSR. A small band of other scientists, and particularly a group of highly vocal younger scientists based in Moscow, urged the new Russian government to "reform" the academy structure. They advocated replacing the elderly leadership—who, they argued, was committed to Soviet-style central control over scientific research—with a new generation of scientific leaders who would promote decentralization of authority to the more than 400 academy research institutions located in Russia. However, their scheme for breaking up the academy was not well developed, and it commanded little support within the government or among the vast majority of scientists themselves.

In late 1991, before the breakup of the Soviet Union, Boris Yeltsin, who was then president of the Russian republic, decided to establish a new Russian Academy of Sciences for the Russian republic; it would operate in parallel with the Soviet Academy. Eighty well-known Russian scientists became its first members. About a month later, however, the Soviet Union collapsed and Yeltsin became president of the Russian Federation. The task that followed was to sort out the relationship between the new academy and the Soviet academy, which had a much larger Russian membership and a heritage that dated back to Peter the Great. The regional problem was quickly addressed, with each new state outside Russia simply taking responsibility for the local academy that had been affiliated with the Soviet academy and then having to find the resources to keep its functioning. But to this day, the financial problems faced by the new states continue, largely immune to resolution.

In Russia, after several weeks of acrimonious debate reported continually in the press and of street protests in Moscow by the advocates of reform of science, Yeltsin decreed that the new Russian Academy of Sciences would be folded into the ASUSSR to form the permanent Russian Academy of Sciences. The outcries for reform, including calls for transforming whatever academy emerged into another type of institution, were quickly muted. Within a few days, the signs on the academy doors were modified, a new academy president with close ties to President Yeltsin was installed, and research activities in Russia continued largely as in the past except that financial support from the government decreased precipitously.[1]

Meanwhile, some scientists from within and outside the ranks of the Russian Academy of Sciences, based primarily in Moscow, were not satisfied with the direction the permanent Russian Academy of Sciences was taking. They decided to establish new academies that would promote their personal interests more effectively (e.g., academies of engineering, aviation, informatics, natural sciences). In time, more than 30 such academies were established, at least on paper. To add to the confusion for foreigners, many old and new educational institutions also include in their names the word academy. But the Russian government strongly backed the Russian Academy of Sciences as the scientific leader of the country. Together with the long-standing Academy of Medical Sciences and Academy of Agricultural Sciences, and their dozens of research institutes, the three academies were far and away the scientific academies with the most stature within the country.

Even though the ASUSSR had employed only about 10 percent of the nation's researchers, the academy system was widely considered to be the

---

[1]Author interview in September 2003 with an academician who was serving as one of the vice presidents of the Soviet Academy of Sciences in 1991.

centerpiece of the nation's scientific effort. And especially for civilian science, the academies had always received priority for personnel, facilities, and equipment. As the economic crisis made itself felt throughout the new Russia, the financial and physical conditions of the RAS and its institutes declined rapidly, although they were in better shape than most university laboratories and research institutes of the ministries. Many prominent American scientists with close ties to Russian investigators within the academy system began calling for dramatic actions by the U.S. government and by U.S. private sector organizations to "save Russian science." They saw a widening stream of excellent scientists emigrating to the United States and other Western countries from Russia. They noted a dramatic decline in research publications from Russian institutions and the dwindling number of Russian participants in international scientific conferences. And they received a barrage of pleas from Russia for humanitarian assistance simply to keep the families of scientists fed and clothed.

As for research within other organizations, many of the institutes of the RAS had ties to higher education institutions. Some of these higher education institutions were important as independent and collaborating research centers.[2] As for the hundreds of applied research institutes of the former Soviet ministries, the most impressive research centers were associated with the Ministry of Atomic Energy, the Space Agency, and the Ministry of Defense, while almost all of the traditionally "civilian-oriented" centers were on the decline.

In the United States, the National Academy of Sciences now found itself with a new partner—the Russian Academy of Sciences—which had inherited the international responsibilities as well as the financial liabilities and physical assets within Russia of the Soviet Academy of Sciences. In late 1991, pleas for Western assistance began to be heard increasingly from leading members of the newly minted academy. The NAS, in close consultation with those U.S. government departments and agencies interested in main-

---

[2]Every year the Ministry of Education rank orders by quality the 700 higher education institutions, using criteria that have not been made public. At the top of the list are universities that have strong ties with RAS institutes and are well known in the West: Moscow State University, St. Petersburg State University, Moscow Physical Technical Institute, Tomsk State University, Kazan State University, Moscow State Technical University (Bauman), St. Petersburg State Technical University, Russian State Oil and Gas Academy (Gubkin), and Tomsk Polytechnical University (see RAS, 2002: 86). A detailed discussion of linkages between academy and educational institutions is set forth in *Integration of Science and Higher Education in Russia* (2001).

taining a strong Russian civilian technical base while preventing the proliferation of weapons from Russia, quickly mobilized resources to begin to develop practical measures for responding to Russia's financial crisis. The response had two thrusts: first, steps to help alleviate the country's general economic crisis and, second, programs to address the calls for immediate help to preserve the country's scientific capabilities.

WEATHERING THE ECONOMIC CRISIS

Although the NAS and the ASUSSR had initiated a series of dialogues in the 1980s on Soviet problems in moving toward a market economy, the Soviet economic crisis had worsened to the point that discussions of specific policies and action programs rather than general dialogues held greater priority for the Russians. The U.S. and other Western governments were considering a variety of multilateral and bilateral assistance programs, but they were hesitant to commit resources until the macroeconomic framework within Russia was to their liking. Consequently, much of the early response from the West was in the form of policy advice from Western economists, many of whom differed in their concepts of what was best for Russia. Within this context, the NAS, acting through the National Research Council, which increasingly drew on members of the National Academy of Engineering and the Institute of Medicine as well as those of the NAS, launched several efforts in the early 1990s designed to provide an improved framework for substantial Western investments in revitalizing an economy on the decline.

These first interacademy efforts centered on "defense conversion"—that is, finding opportunities for Russian defense enterprises to develop and manufacture products for the civilian market. Toward this end, an NRC committee undertook case studies of the commercial potential at two military aviation enterprises in Saratov—Tantal and the Saratov Electromechanical Production Organization (SEPO). Among the recommendations developed was the call for establishment of a U.S.-Russian fund that would serve as a loan mechanism for Western investments in defense conversion activities, with U.S. government assurances that American investors would not incur large losses should their investments turn sour. A related effort centered on documenting lessons learned from a comparative review of regional defense conversion plans in different areas of Russia (NRC, 1993c, 1993d).

Closely related to this interacademy focus on defense conversion was another project on dual-use technologies. It concentrated at first on the activities at defense firms in Perm that had always worked behind a veil of secrecy. In addition to pondering the difficulties encountered in trying to attract investors to finance civilian-oriented activities at these firms, the project looked at the export control aspects of marketing dual-use technologies internationally. During the 1980s, NRC expert committees had conducted several studies of export control issues that provided helpful background for this project (NRC, 1987b, 1991b). The project culminated in a joint report of the two academies that set forth a set of guidelines for balancing the commercial and export control objectives of Russian conversion programs. The report was particularly helpful to several Russian legislators during the 1993 debates of dual-use issues in the lower house of the Russian parliament, known at that time as the Supreme Soviet (NRC, 1994).

A third interacademy effort was the continuation of the exchange of young researchers in economics. At meetings in Moscow and Leningrad in mid-1991, young specialists from the two countries prepared joint papers and set the stage for sustained cooperation on a direct specialist-to-specialist basis. The topics of continuing interest included internal currency markets, price reform, enterprise reform, and environmental economics (NRC, 1991c: 12).

These interacademy activities in the field of economics had little immediate influence on policies in Russia or the United States because of the many larger efforts being carried out under the auspices of various Western governments. However, the dialogues sponsored by the academies attracted important specialists from both countries who subsequently played a significant role in developing the intellectual framework for the never-ending debates about the economics of a country in transition. Thus, the interacademy program clearly had an indirect "educational" impact on the economic policies designed and implemented in Russia during the 1990s.

SAVING RUSSIAN SCIENCE

At the request of the White House Office of Science and Technology Policy (OSTP), in early 1992 the National Academy of Sciences brought together over 100 policy officials and technology specialists from the American science and technology community to consider how to energize Russia's rapidly declining research capability. They considered (1) the decline in pay for Russian weapons scientists and engineers who might be tempted to look to unreliable countries for financial support; (2) the Russian government's dramatic cutting of support for basic research programs; (3) the difficulties encountered in trying to commercialize Russian technology; and (4) the lack of significant programs of interdisciplinary, problem-oriented research. Recommendations in each of these areas were included in the report of the conference (NRC, 1992d).

Several suggestions were eventually incorporated into intergovernmental and private sector programs, with advocacy by the NAS an important factor in encouraging their acceptance. The recommendations highlighted, for example:

• the importance of adopting a broad definition of weapons scientists who would be eligible for financial support through the soon-to-be-established International Science and Technology Center in Moscow. (In short order, the center adopted this approach.)

• the opportunities for easily dispensing new funding for cooperation with Russian colleagues through the extramural programs of the National Institutes of Health (NIH), National Science Foundation, and Department of Defense. (This concept was almost immediately put in place by several agencies.)

• endorsement of a congressional initiative to establish a research foundation for support of science in the former Soviet Union. (Several years later, Congress established the U.S. Civilian Research and Development Foundation.)

Other perennial issues that were considered, but that garnered few new ideas for their resolution, included the inadequate legal framework for intellectual property rights in Russia, the stifling effect of taxes on grants for scientific research, and the need to refurbish Russian laboratories on a massive scale.

The 1992 gathering of American specialists was only one in a series of meetings that set out to "save Russian science." However, even at this early date many leading Russian scientists found this objective far too broad and began calling for steps to save world-class "schools of Russian science." Subsequent interacademy meetings attended by RAS leaders and significant Russian government officials, as well as leading American officials and representatives of professional societies, were of special importance in such an effort. These meetings were held in Washington and Moscow during 1992 and 1993. Among the topics of special concern were the brain drain from Russia and within Russia; ways to stimulate and retain the interest of Russian youth in science, including the broadening of access to computers in schools; and international concerns over global ecology and energy issues.

These topics had been raised in U.S.-Soviet dialogues in previous years, but with the Russian economy in a shambles there seemed to be a more receptive audience among funders searching for program opportunities. Also of continuing concern was the stagnant state of social science research in Russia—the result of most of the accomplished social scientists being called to government service. Finally, the NRC developed an ambitious plan to analyze the mobility of Russian scientists within the country, but, lacking a financial sponsor, the plan was destined to remain in a file drawer (NRC, 1992c: 4).

In late 1992 the NRC organized another significant conference in Washington that developed a framework for a proposed action plan that cut across many topics (NRC, 1993e). These topics included:

• organizing a new program of competitively awarded research grants, with the funds to be transferred directly into the hands of the selected Russian scientists
• collecting books, journals, and other material to help restock depleted Russian scientific libraries
• rebuilding the human and physical infrastructures needed to support scientific research
• preserving unique specimen collections and data sets in various fields
• organizing outreach programs in Russia to stimulate public support for science
• devising a mechanism to coordinate Western assistance programs that could help Russian science and technology activities.

The NRC then turned again to the poor state of social sciences in Russia. At another 1992 meeting, leading American and Russian specialists urged government action to help maintain the libraries of Russia, to upgrade them with computer systems and reference material, and to link them together electronically. The group also pointed to the need to identify and preserve the many data sets scattered throughout the country. And they called for greater opportunities for researchers to spend time in the Unites States, where they could develop model curricula for use by Russian universities (NRC, 1992a).

Also during 1992, the NAS supported the initial deliberations organized by businessman and philanthropist George Soros that eventually led to the establishment of the International Science Foundation. The foundation was designed to provide temporary financial support for selected Russian researchers. The NAS leadership actively participated in development of this initiative, and the NRC played a significant peer review role in ensuring the scientific integrity of the life sciences program that emerged (Dezhina, 1999).

The NAS also reached out to counterpart organizations in Europe, which shared concerns about the decline of science in Russia. At meetings hosted in 1995 and 1996 in London by the Royal Society and in Moscow and Paris by the Organization for Economic Cooperation and Development (OECD), NAS representatives helped to chart possible paths for collaborative programs. At these meetings it became clear that the NAS was well ahead of European academies in putting in place its own programs and in encouraging the government to pay greater attention to the plight of Russian scientists and engineers (NRC, 1992b: 3).

REORIENTING THE INTERACADEMY EXCHANGE PROGRAM

In 1992 the NAS and the RAS entered into a new interacademy agreement that significantly changed the approach to exchanges of individual scientists. For more than 30 years, individual exchanges had been based on nominations of scientists by the sending academy. The receiving academy would then attempt to place the nominees at appropriate research facilities for periods of two weeks to one year. As noted earlier, during the 1980s the NAS had tried to encourage nominees to obtain invitations for visits from colleagues in the other country, but this approach was not working well. In fact, most of the nominees did not have invitations from prospective hosts. The selection of some Soviet nominees had continued to be based more on the relationship of the nominees with the ASUSSR bureaucracy rather than on scientific merit. Also, American hosts simply did not know many of the nominees and agreed to receive them as a courtesy to the NAS rather than from a desire to establish genuine collaboration (NRC, 1991a: 1).

The new system under which Russian scientists would travel to the United States was simple, at least from Washington's perspective. The host American scientist would invite a Russian colleague to visit and then apply to the NAS for financial support. The NAS would decide which applications to support on the basis of merit, knowing that the hosts—who were the applicants—were genuinely interested in receiving the visitors. As for American travelers to Russia, the interested American would apply to the NAS with an invitation from a Russian colleague in hand, and again the participants would be selected on the basis of merit. In both cases, the American scientist would make all the administrative arrangements, with the NAS simply providing a travel grant. At first, this change caused some consternation on the Russian side, because the RAS had its own list of Russians it wanted to send to the United States. But soon the new approach was accepted, and the exchange program, which now is better described as a "grants" program, has worked well since the change, which is described in more detail in Chapter 5.

Meanwhile, the NAS and the RAS leaderships established an ad hoc interacademy panel to recommend new modes of cooperation. The suggestions of the panel seemed sound, but most were not implemented because of a lack of financial sponsors. These suggestions included the establishment of electronic clearinghouses in the two countries to track cooperative projects, and thereby avoid unnecessary duplication, and an emphasis on exchanges in areas in which Russian institutions had different but complementary capabilities such as high-temperature superconductors, the physics and chemistry of fullerenes, and energy conservation. One suggestion, which was adopted many years later on a limited basis, called for inviting Russian specialists to serve as members of NRC study panels (NRC, 1993a: 1).

By 1994 the NAS and RAS had considered a broad array of policy and program initiatives by the two governments, by the academies, and by other institutions that could be helpful as Russia began to realign its political structures and to move toward a market-oriented economy. The time had come for launching new projects of interest to members of the two academies, and the remaining chapters of this report address some of the most important interacademy projects during the past decade.

These projects were, however, affected by two important changes in U.S.-Russian scientific cooperation during the early 1990s as the Russian economy continued its downward spiral and as Western governments adopted a foreign assistance mentality in dealing with Russia. First, almost all American organizations began paying the expenses associated with Russian visitors traveling to the United States in addition to covering the expenses for American visitors to Russia. The principal exception was official delegations sent by the Russian government to the United States—these delegations paid their own way, using Russian government funds. In most cases, then, the NRC has had to shoulder the task of raising larger sums of money than in earlier years when the costs of cooperative projects were shared, and this practice of the NRC carrying most of the financial burden continues to this day.

Second, there was a sharp decline in the number of Russian specialists with international experience who were in a position to arrange cooperative programs for academy institutions. Some of their most skilled colleagues had taken more lucrative positions in the private sector and were not replaced. At the same time, so many Western organizations were suddenly interested in arranging cooperative projects that the Russian gatekeepers were overwhelmed. They preferred to put at the head of the line the deal that appeared to be the most financially rewarding for themselves. Fortunately, the activities of the NAS rooted in formal interacademy agreements continued with few disruptions, but occasionally projects were delayed as other more lucrative arrangements received new attention in Moscow.

In short, just as life in Russia was rapidly changing, the character of U.S.-Russian cooperation was undergoing a major transformation as well. But as the opportunities for cooperation expanded dramatically, there was a significant danger that the quality of programs would decline. In some instances, Western enthusiasm to visit previously isolated geographic regions in Russia, to walk through closed facilities, and to drink vodka with new acquaintances with innovative achievements in their résumés pushed considerations of quality and the potential impact of new exchange activities into the background. Perhaps the greatest contribution of the NAS to science during this period of transition was its steadfast determination to demonstrate that quality still mattered in cooperative undertakings. Other, less visible organizations also consistently stressed scientific integrity, but some organizations on both sides of the ocean seemed more interested in having scientific "events" than in advancing science.

4

National Security Issues and a Wider Agenda for Cooperation

The diffusion of military technologies into the civilian sector was of a "semi-military" character. Dual-purpose technologies were then developed in the civil sector that could be used for military purposes.
Policy Report, Russian Ministry of Science and Technology, 1993

The new Russia emerged in a confrontational and sometimes violent environment. In 1991 Russian president Boris Yeltsin mounted an army tank in the center of Moscow to defy Soviet authorities, and Soviet president Mikhail Gorbachev was apprehended by communist reactionaries at his Black Sea retreat only to be rescued by Yeltsin's forces. In 1992 fires and repeated demonstrations erupted on the streets of several Russian cities. The year 1993 saw, among other things, a Russian tank unit shell the parliament building in Moscow, which was occupied by defiant and armed legislators. In the years that followed, leading political figures were assassinated, violence ripped through ethnic enclaves in various parts of the country, and murder became a favorite tactic of robber barons seeking control of the country's financial assets.

Meanwhile, the protracted conflict in the Russian republic of Chechnya triggered hundreds of kidnappings and dozens of bombings of buildings and vehicles in some Russian cities. In October 2002 Chechen rebels seized a Moscow theater, and 130 of the 800 hostages died, most from an anesthetic gas released by Russian security forces at a high dosage level to incapacitate the Chechen militants. At the same time, the security forces shot and killed all 40 Chechen rebels, and they succeeded in preventing detonation of any of the explosives strapped to the female Chechen hostage-takers and emplaced on the support structure of the theater. More than 1,000 people inside and outside the theater could have been killed had not the raid of the theater succeeded.[1] Although governments worldwide have been dismayed by the brutal conduct of Russian army troops during the seemingly merciless operations in Chechnya, the Russian population remains generally supportive of "whatever it takes" to track down the hit-and-run "terrorists" hiding in the North Caucasus.

It is no wonder, then, that Western governments continue to be concerned about the stability of a country that is a repository of huge stockpiles of nuclear weapons, discarded but functional chemical weapons, and microbes and viruses that could be used in biological weapons. Even if the Russian government is committed to safeguarding all dangerous weapons and materials in its inventories, both from internal dissidents and from international pirates, the high-powered weaponry in Russia remains a tempting target for rogue states and terrorist groups with access to large sums of money. Rumors have circulated for years that some nuclear weapons or nuclear materials from Soviet stockpiles were obtained by international groups hostile to Western interests in the early 1990s, when the well-financed Soviet security system gave way to a dysfunctional Russian system. Even today, Western experts remain concerned about the weaknesses in the improved Russian security systems and the ability of those systems to contain all dangerous weapons and materials.

The tens of thousands of Russian weapons scientists, engineers, and technicians with special knowledge about weapons of mass destruction pose a particular problem. Having lost most of the economic privileges accorded Soviet weaponeers, and indeed in many cases having lost their jobs, these specialists are potential "know-how" targets for shadowy groups determined to develop and build their own advanced weaponry. Even though there is no evidence that Russian specialists have been effectively recruited from abroad as channels of sensitive information or as accomplices in schemes to steal lethal materials or weapon components, they continue to be at the top of the list of potential risks to the security interests of Western nations.[2]

Beginning in 1991, the U.S. government adopted programs—most notably, the Cooperative Threat Reduction, or Nunn-Lugar, program—to help

---

[1] For details on this hostage incident, see Popova (2002).
[2] For a detailed discussion of the weapon scientists, see Schweitzer (1996).

reduce and contain weapons material and expertise in Russia. Over the past decade, the U.S. Congress has appropriated more than $7 billion for nonproliferation efforts, including not only efforts by the Department of Defense, but also a dozen cooperative programs managed by the Department of Energy (DOE) and several programs of the Department of State.[3] Soon, American personnel began working on joint projects at some of the most sensitive facilities in Russia. As access to such facilities increased, the precarious economic conditions throughout the Russian weapons complex became more obvious to Western visitors. They recognized the need for aggressive cooperation, with money earmarked for Russian participants to help stabilize the situation.

Against this background, and with the encouragement of the Departments of State, Defense, and Energy, the National Research Council began in the mid-1990s to launch a series of studies and related efforts to address threats to the international security interests of the United States stemming from the breakup of the Soviet Union. In a sense, any type of NRC project undertaken with the participation of Russian institutions could be considered an international security activity, given the high security stakes involved in the U.S.-Russian relationship. But for the purposes of this report, only those projects that have direct linkages to traditional concerns about access to dangerous material, weapons, and other destructive devices are placed under the heading of national security. In all of these projects, the Russian Academy of Sciences has played a role—either as a full partner or as a facilitator of NRC interactions with Russian ministries and other organizations within and outside the academy system.

As noted in the preface to this report, the activities of the parallel U.S. and Russian academy Committees on International Security and Arms Control are not considered in this discussion. For many years, they have addressed some of the issues described here as well as other bilateral security concerns, thereby giving further weight to the commitment of the RAS and the U.S. National Academies to work in the national security arena.

PROTECTING NUCLEAR MATERIAL

Throughout Russia, vast quantities of plutonium and highly enriched uranium suitable for use in weapons are kept in hundreds of buildings at

---

[3] Source for $7 billion is the Russian-American Nuclear Security Advisory Committee, August 2003.

dozens of facilities. It is not surprising, then, that the inadequacies of the security systems surrounding this material became an early object of interest for security-oriented studies carried out by the NRC. Two NRC reviews of the Department of Energy's materials protection, control, and accounting (MPC&A) program were conducted in the late 1990s. The conclusions of both studies strongly supported continuation of the program being carried out by the DOE in cooperation with Russia's Ministry of Atomic Energy to upgrade the capabilities of nuclear facilities in Russia. Other recommendations of the second study, which closely paralleled those of the first study, called for (1) reviewing priorities to address important vulnerabilities, (2) "indigenizing" MPC&A capabilities, (3) reducing impediments to effective cooperation, and (4) improving the management of U.S. personnel and financial resources (NRC, 1997a, 1999).

Beginning with President Bill Clinton himself, many officials within the executive branch were interested in the details of the conclusions of these reports.[4] On Capitol Hill, interest also ran high in view of widespread skepticism about the effectiveness of the U.S. government's effort and, in particular, of worries that the MPC&A program had become a funnel for channeling most of the appropriated funds to the coffers of the participating DOE laboratories rather than to Russian organizations. The NRC studies probably were helpful in reassuring congressional critics of the program that investments in MPC&A upgrades in Russia were indeed worthwhile and that shortcomings in the management of the program could and should be overcome, including improvements in determining the portion of the appropriated funds that should be spent in Russia. In any event, Congress increased the budget for the MPC&A program significantly in the year after completion of each of the studies, and senior DOE specialists continue to use recommendations in the reports as reminders of areas needing greater attention, even in 2003.[5]

An important conclusion of each study was the need to encourage the Russian government and Russian institutions with nuclear material to demonstrate explicit commitments to maintaining the MPC&A upgrades once U.S. financial support ended. The DOE should therefore orient its program toward activities that would ease the transition to sustainability. This orientation should include greater use of locally manufactured equipment,

---

[4] Discussions with National Security Council staff, March 1997.

[5] Briefing material on MPC&A program presented by DOE officials at a meeting in Oak Ridge, Tennessee, in September 2003.

more reliance on less energy-intensive approaches than those adopted at Western facilities, and increased consolidation of weapons-grade material at fewer locations, for example. Even though such commitments to sustainability were considered essential, there seemed to be only a limited response in Washington and Moscow to the recommendations. Therefore, in 2003 the NRC undertook a third study of MPC&A upgrades in Russia that looked exclusively at the sustainability of the upgraded systems into the indefinite future. Perhaps a better descriptor of the thrust of the study is "indigenization" of MPC&A responsibilities. Some of the approaches that were highlighted in the earlier studies and that are among those being considered in this study are the following:

• Increase the percentage of available funding directed to financing activities of Russian organizations, with a steadily declining percentage directed to supporting U.S. participants in the program.

• Expand efforts to utilize Russian equipment and services whenever possible and to encourage Russian enterprises and institutes to increase capabilities to provide high-quality equipment and associated warranties and services.

• Use Russian specialists from institutions with well-developed MPC&A capabilities to replace some U.S. members of teams at Russian institutions with less-developed capabilities.

• Rely increasingly on Russian specialists to replace U.S. specialists in presenting MPC&A training programs in Russia.

• Encourage the Moscow Physics and Engineering Institute to increase student participation (and its income resulting from tuition payments) in its security-oriented courses by offering an industrial security specialization alongside its MPC&A specialization.

• Give greater attention, in both training and implementation activities, to developing personal commitments on the part of Russian managers, specialists, and guard forces to fulfill their responsibilities for ensuring the proper functioning of MPC&A systems (NRC, 1999).

In a related activity, in 2002 the presidents of the National Academy of Sciences, National Academy of Engineering, and Institute of Medicine, together with the president of the Russian Academy of Sciences, decided to place a greater emphasis on nuclear nonproliferation issues in the interacademy program, and particularly protection of nuclear material. This emphasis was underscored in a joint statement by the four institutions (see

Appendix E). A special interacademy working group was established to develop both recommendations to governments and project proposals in the general area of nuclear security (see Box 4-1 for the initial conclusions of the working group). In its recommendations, the group continued to emphasize MPC&A as an important area for cooperation. The subjects of two new program initiatives for immediate implementation were (1) impediments to cooperation in nuclear areas (e.g., delays in issuance of visas, limitations on access to sensitive facilities, assumption of liability for injuries suffered during cooperative projects), and (2) best practices worldwide in establishing and operating MPC&A systems, including the use of remote monitoring.

CONTROLLING EXPORTS OF NUCLEAR AND
OTHER DANGEROUS MATERIALS

In any country, programs designed to ensure that nuclear materials will not be stolen or diverted should be closely linked to effective systems to control the commercial exports of nuclear and other dangerous items within the framework of international export control regimes. As noted in Chapter 3, in the 1980s the NRC had conducted several studies of U.S. policies for controlling the export of sensitive items and related technical data. The new challenge for the NRC in 1996 was to assess the effectiveness of U.S.-Russian cooperative programs designed to improve export control policies and systems in Russia.

In Soviet times, the government had effectively controlled exports of almost all materials—military and civilian. But during the 1990s, government officials were preoccupied with promoting profit-oriented activities rather than restricting exports that could generate income. Immediately after the rebirth of Russia, the borders of the new state could best be described as porous. There was a rush within Russia and abroad to take out of Russia valuable equipment, materials, and technical information, including sensitive items. Western governments eagerly participated in what became known as Russian technology bazaars, assuming they would be able to acquire advanced technologies without the constraints of patents, and that they would even learn valuable defense secrets. Russian entrepreneurs were convinced that they could reap profits from trading with the country's most prized assets. The problem was not one of reinventing Soviet export control systems to prevent the unbridled proliferation of dangerous items, but rather one of updating the well-developed existing systems and, most important,

BOX 4-1 Recommendations for Interacademy Action in the Field of Nuclear Security, 2002

The following recommendations were developed by a joint committee of the U.S. and Russian academies and transmitted to the leaderships of the academies and the appropriate government officials.

Recommendations to Governments

• Appoint a single high-level official in each government to ensure that continuing attention is paid to diminishing the obstacles to and exploiting the opportunities for bilateral cooperation on nuclear nonproliferation and counterterrorism.

• Increase the priority of and resources for a "security first" agenda for reducing the risks from stocks of highly enriched uranium (HEU) and separated plutonium by consolidating material at fewer locations, accelerating the blend-down of HEU to levels that do not pose a threat, and minimizing use of HEU in research reactors.

• Expand cooperation in dismantling Russian general-purpose nuclear submarines.

• Give higher priority to information and education efforts on the risks of handling nuclear materials improperly.

Recommendations for Interacademy Projects

• Overcome impediments (e.g., visa, access, and liability problems) to U.S.-Russian cooperation in nuclear nonproliferation activities.

• Identify best practices for materials protection, control, and accountability (MPC&A) worldwide, including remote monitoring techniques.

• Assess cooperative approaches to promote conversion of research reactors from highly enriched uranium to low-enriched uranium.

• Develop a road map for Russian general-purpose submarine dismantlement and management of naval spent fuel.

• Assess cooperation in U.S.-Russian MPC&A programs.

Source: Adapted from "Letter Report from the Co-Chairs of the Joint Committee on Nuclear Nonproliferation," December 4, 2002, released by the National Academies and the Russian Academy of Sciences.

enforcing regulations that were in place but were being circumscribed by Russian industrial oligarchs and other persons of influence.[6]

As in its assessment of the MPC&A program, the NRC concluded in its 1996–1997 export control study that cooperative programs to strengthen export control procedures (led by the Department of Commerce on the U.S. side) were important and were helping to tighten Russian export controls. Probably the most significant contribution of the NRC study was its advocacy of greater attention to establishing export control competence and commitment within large Russian industrial firms, which were the birthplaces and repositories of most of the items of concern. Such a recommendation was, however, contrary to the immediate export interests of many high-tech enterprises in Russia, which preferred to plead ignorance about export controls when questioned as to their policies. Nevertheless, in time, the concept took hold in Russia, and, just as in the United States, individual companies eventually began to take seriously their responsibilities to comply with the law (NRC, 1997a: 85–117).

During the study, the permissiveness of the international regimes for controlling exports of items with military significance became evident. Although Russia was improving its procedures to help ensure that the government was aware of all proposed exports, the government's decisions on the appropriateness of many types of exports were subject to only limited international constraints. For example, the U.S. government has been particularly concerned about exports of nuclear-related items from Russia to Iran. But within the international nuclear control regime, Russia has wide latitude to export items for civilian nuclear power plants that it considers appropriate, and Russia's views on appropriateness—driven in large measure by financial considerations—differ markedly from U.S. views.

The 1997 NRC study report highlights the importance of controlling technical data associated with sensitive technologies (NRC, 1997a). In recent years, control or lack of control of technical data has become a significant issue in U.S.-Russian relations. Protection of technical data had for many years affected the issuance of U.S. visas for meetings attended by American specialists who might enter into discussions of technical data that should be subject to export control. But during the 1990s, these concerns seemed to fade in importance as other issues dominated the U.S.-Russian relationship. Then, at the turn of the century, security-related requirements

---

[6] For a detailed discussion of the control of exports during this period, see NRC (1997a: Chap. 5).

for issuance of U.S. visas were tightened because of both concerns about terrorism and a revived interest in preventing the proliferation of sensitive data. At home, the National Academy of Sciences, National Academy of Engineering, and Institute of Medicine have expressed concern to the U.S. government about the impact on scientific research of restrictive policies that limit the international exchange of research concepts and results, a perennial issue that dates back several decades.[7] Among many Russians, and indeed among American specialists, there is often confusion about the overlap among controlled technical data, proprietary information, and classified information, and this overlap will undoubtedly produce recurrent uncertainties for specialists involved in high-tech ventures.

THE WAYWARD WEAPONEERS

The technical prowess of Russian weaponeers who no longer had stable employment and paychecks that covered their essential costs of living became of special concern to the NRC in the early 1990s. Having supported the efforts within the Department of State and on Capitol Hill to establish the International Science and Technology Center (ISTC) in Moscow as a mechanism for financing the redirection of the research efforts of scientists from military to civilian endeavors, the NRC welcomed the opportunity offered by the Department of Defense (DOD) to review the initial research programs and operating procedures of the ISTC, which by that time had committed over $50 million to research projects in Russia. The NRC study released in 1996 concluded that although the ISTC had been operating for only two years, it was a low-cost, noncontroversial program that was already providing jobs—full time or part time—for 12,000 scientists. The report urged that funding of the ISTC be increased (NRC, 1996a). It also advocated greater attention to biological and chemical issues, involvement of the private sector, and communications projects. By 2003 new programs in all three areas were in place at the ISTC, and the commitment of funds by the founding governments and their partners had exceeded $500 million (ISTC, 2003).

In the mid-1990s, several other U.S. government programs were established to help reduce the likelihood that scientists and engineers with specialized knowledge would be tempted to look to foreign traders with unreliable clients for financial support. Of special interest were the new programs

---

[7] See, for example, NRC (2002a).

of the DOE and the Department of State, which combined government resources with financial contributions from Western companies that could profit from working with former Soviet weaponeers. Given its strong connections with the U.S. private sector, the NRC was in an excellent position to suggest how the interests of the Russian scientists, the U.S. government, and private companies could be effectively combined. In 1996 and again in 2002 the NRC initiated in-depth assessments in this important arena, with a focus on the biological sciences as discussed in the next section.

REDIRECTING RUSSIAN BIOLOGICAL EXPERTISE FROM MILITARY TO CIVILIAN PURSUITS

In 1996 the leadership of the NRC and DOD officials held discussions over many months about Russian capabilities with implications for biological weaponry and biological terrorism. At the center of the discussions were proposed steps that could help provide assurance that the Russian government had abandoned its offensive biological weapons program, that it was complying with the Biological Weapons Convention, and that it was not providing militarily sensitive materials or expertise to states of proliferation concern. Finally, the DOD decided to support an NRC effort to help chart a course for engaging former biological defense scientists from Russia and several other former Soviet republics in a cooperative program that would provide assurances in these areas.

The NRC project followed two tracks. First, American specialists held intensive consultations with a large number of Russian officials and scientists on the feasibility of cooperative research projects in the biological sciences and their likely contribution to transparency throughout the former Soviet weapons complex. Second, the NRC convinced the DOD to finance eight pilot cooperative research projects to demonstrate the opportunities and problems encountered in launching cooperative projects to address issues pertinent to Russian facilities that had previously been off-limits.

Most of the consultations on research opportunities, transparency, and related issues, as well as the development and conduct of the eight pilot projects, were centered on institutes in the Biopreparat complex. At the height of its activities in the 1980s, Biopreparat had employed thousands of highly skilled scientists and engineers dedicated to supporting the Soviet bioweapons program. The complex included dozens of facilities designed for research on highly dangerous pathogens with clear military applicability and several facilities built to produce large quantities of agents for biological weapons. Although the research institutes had been active for more than 20 years, the production facilities had remained in a stand-by mode. The NRC-led consultations involved a variety of meetings, most notably a scientific symposium in Kirov where the principal Russian military biological research facility is located, a workshop organized by Biopreparat near Moscow, and lengthy discussions during visits to several Biopreparat institutes. The NRC specialists gained unusual insights into past Russian activities and the future aspirations of Russian specialists. Meanwhile, the pilot projects were carried out with minimal difficulty or delay (see Box 4-2 for a list of the projects).

BOX 4-2 Pilot Projects Initiated by the National Research Council and Financed by the Department of Defense

The following pilot projects were carried out from 1997 to 1998, with funds committed to Russian institutions.

At the State Research Center for Virology and Biotechnology "Vector," Koltsovo:

- study of the prevalence, genotype distribution, and molecular variability of isolates of hepatitis C virus in the Asian part of Russia; $55,000; principal investigator, Sergei Netesov; collaborator, Elizabeth Robertson, CDC; ISTC 883
- study of the monkeypox virus genome; $55,000; principal investigator, Sergei Shchelkunov; collaborators, Peter Jahrling, USAMRIID, and Joseph Esposito, CDC; ISTC 884
- study of the genetic and serological diversity of hanta viruses in the Asian part of Russia; $55,000; principal investigator, Lyudmilla Yashina; collaborators, Connie Schmaljohn, USAMRID, and Stuart Nichol, CDC; ISTC 805
- development of advanced diagnostic kit of opistheorchiasis in human patients; $55,000; principal investigator, Valery Loktev; collaborator, Victor Tsang, CDC; ISTC 691
- experimental studies of antiviral activities of glycyrrhyzic acid derivatives against Marburg, Ebola, and human immunodeficiency virus; $51,683; principal investigator, Andrei Pokrovsky; collaborator, John Huggins, USAMRIID; ISTC 1198

(continued)

BOX 4-2 *(continued)*

At the State Research Center for Applied Microbiology, Obolensk:

- molecular-biological and immunochemical analysis of clinical strains of tuberculosis and mycobacteriosis; $138,000; principal investigator, Igor Shemyakin; collaborator, Thomas Shinnick, CDC; ISTC 810
  - investigation of the immunological effectiveness of delivery *in vivo of the Brucella* main outer membrane protein by the anthrax toxin components; $61,500; principal investigator, Anatoly Noskov; collaborators, John Collier, Harvard University, and Arthur Friedlander, USAMRIID; ISTC 919
  - monitoring of anthrax; $55,000; principal investigator, Nikolai Staritsin; collaborator, Arthur Friedlander, USAMRIID; ISTC 1215

Note: CDC = U.S. Centers for Disease Control and Prevention; DOD = U.S. Department of Defense, ISTC = International Science and Technology Center (Russia); USAMRIID = U.S. Army Medical Research Institute of Infectious Diseases.
    The following funds were committed to U.S. collaborating institutions: CDC, $47,000; USAMRIID, $20,000; Harvard University, $9,000.
Source: NAS/IOM/NRC (1997: 1).

The NAS/IOM/NRC report issued in 1997 called for a long-term, multimillion-dollar annual effort by the DOD to engage American and Russian specialists in research efforts at institutes formerly involved in the Soviet defense program on topics that would be of interest to the DOD's biodefense efforts. In summary, the report stated:

After extensive consultations with key Russian officials and scientific leaders and drawing on the experience gained through the initiation of pilot projects at two Russian facilities to investigate the practical aspects of cooperation, the National Academy of Sciences Committee on U.S.-Russian Cooperation on Dangerous Pathogens recommends a five-year *Pathogens Initiative,* followed by a second phase of sustained joint U.S.-Russian research and related efforts. The program will support collaboration on the epidemiology, prevention, diagnosis, and therapy of diseases associated with dangerous pathogens that pose serious public health threats, as well as related fundamental research. The *Pathogens Ini-*

tiative will engage a substantial number of highly qualified specialists from the former Soviet biological weapons complex and will serve important U.S. national security and public health goals. (NAS/IOM/NRC, 1997: 1)

Within a year of release of the NAS/IOM/NRC report, the DOD had committed to a substantial program of cooperative engagement with Russian research institutes, largely along the lines suggested by the NRC (NAS/IOM/NRC, 1997).

Identified in the report were research topics of considerable scientific interest but distant from the biodefense priorities of the DOD. Indeed, most of the topics were in the fields of public health and agriculture. Building on the momentum being developed as the DOD began funding cooperative projects in areas clearly relevant to biodefense, the National Security Council and the Department of State, with support from the DOD, succeeded in convincing both the U.S. Department of Health and Human Services (HHS) and the U.S. Department of Agriculture (USDA), as well as key congressional staff members, that these civilian-oriented departments also should become engaged in biological redirection efforts in Russia. Within two years, each of these departments had annual multimillion-dollar programs in place to support cooperative research projects with former bioweapons scientists in fields that were beyond the mission of the DOD but were directly related to the missions of the two departments. Also, the DOE was able to adapt one of its programs to opportunities for supporting research at former weapons institutions, and the Department of State used some of its funds to expand biological redirection activities supported through the International Science and Technology Center. Clearly, the NRC initiative to work with the DOD had spin-offs and played an important role in the development of cooperative programs in several executive departments.

As the intergovernmental activities have expanded, an NRC committee of experts has continued to play a role in biological redirection activities, providing the DOD with reviews of the scientific merit of collaborative projects proposed by Russian institutes. Among the criteria for judging scientific merit are (1) the scientific significance of each proposed project, (2) the quality and feasibility of the proposed research methodology, (3) the track record of the principal investigator and the supporting research team, and (4) the capability of the host institute to provide support for the project. In addition, the committee is carefully assessing the contribution of each proposed project to transparency in Russia and also has pointed out, where appropriate, when a project might produce dual-use results of value to future weapons programs. The committee has noted that in both of these areas the active participation of an American collaborating scientist in each project is essential, and therefore it has helped to identify such collaborators. The NRC committee is also evaluating the progress being made in projects under way and identifying follow-on activities that make sense.

The NRC experts have reviewed and recommended support for many more projects than the DOD has been willing to fund, largely because of difficulties in closely monitoring research that has dual-use implications in high-hazard Russian facilities. This gap between recommendations and financial support has disappointed some of the project managers and scientists of the unfunded projects. By mid-2003 the DOD had considerable uncommitted funds in the pipeline, and at that time it seemed that some of the good project proposals that had been languishing for attention for months and even years since the NRC reviews would be retrieved and supported.

In 2003 the NRC committee of experts initiated a study on the future of biological research and development activities in Russia, with completion scheduled for mid-2004. The study, supported by the Nuclear Threat Initiative, a U.S. private foundation, is assessing the future of Russian efforts to address public health concerns (issue 1), and particularly the spread of infectious diseases; evolution of a biotechnology industry (issue 2); and the basic and applied research essential to support efforts for improvements in these and related areas (issue 3), as well as the intersections between efforts in these three areas and Western concerns about bioterrorism (issue 4) and bioproliferation (issue 5). This broad examination of the long-term, biology-related capacity of Russia, being carried out in close consultation with Russian specialists, is intended to provide insights in several areas:

• How can the resources that will be available within Russia to support disparate programs under the purview of several Russian ministries and other government-affiliated organizations be used more effectively in addressing the five issue areas?

• How can important but underfunded Russian research and development programs be sustained over the long term, with particular attention to the problems of attracting new researchers to careers in the biosciences while also encouraging outstanding scientists who are in place to continue their careers in Russia?

• How can Russia reestablish a pharmaceutical industry that can begin to reduce dependence on imported vaccines, drugs, and diagnostic kits, and eventually can reach out to foreign markets, initially in the former Soviet states and later in the West? Of special interest is the creation of friendly regulatory and tax regimes for both Russian and foreign investors.

• How can international cooperative programs, motivated by both security and nonsecurity concerns, be more effective in supporting a Russian agenda in each of the five issue areas? The involvement of the U.S. Agency for International Development (USAID), the World Health Organization (WHO), the European Union, and UN agencies, along with organizations supporting biological redirection programs, is of interest. Also, new approaches to the biosciences in Russia that are local, regional, and global both in their organizational structure and in their impact are important.

Several aspects of ongoing U.S. government programs are being considered as well. They include:

• assessing the impacts of bioresearch projects supported by the U.S. government on the research communities of Russia, with attention given to both developing scientific knowledge and broadening transparency at participating institutions
• identifying specific benefits from projects that have been supported
• identifying attractive research areas for emphasis in future cooperative projects.

As noted, a large effort is under way within the U.S. government, supplemented by the activities of other governments, to engage former Soviet bioweaponeers in redirection activities. The NRC will play a limited but significant role in this effort—it will review the scientific significance of projects—but it faces the constant challenge of not unnecessarily duplicating the efforts of others. Of particular concern is the large number of overtures from U.S. organizations to a limited number of overloaded international interlocutors on the Russian side to explore opportunities to work together on scientific problems that have been addressed by many other American specialists on earlier visits to Russia.

COUNTERTERRORISM ON CENTER STAGE

In the late 1990s, NRC staff began informal discussions with RAS officials and Russian specialists on crime and terrorism as potential topics for joint interacademy efforts. At about the same time, the U.S. Congress began enacting special legislation on counterterrorism measures to be taken in the United States. By 2000 the Russian government had established an interagency organizational framework for counterterrorism, headed by the Federal Security Service (FSB), thereby overcoming the principal reason for RAS reluctance to engage in international activities in this field—the RAS simply did not want to be in front of the government. Russian army and security forces had been involved for years in attempting to stymie Chechen activities they considered to be terrorism, but the new mandate for the FSB was broader and included coordination with security services of other former Soviet republics as well as coordination within the Russian Federation.

The academies organized a workshop on high-impact terrorism in June 2001 in Moscow. The attendance on the Russian side exceeded the expectations of the organizers in both countries. Presentations were made by representatives of the Duma, Ministry of Internal Affairs, Minatom, FSB, and Ministry of Interior, as well as by some well-known Russian specialists in fields relevant to counterterrorism. In the audience were numerous representatives of the Ministry of Defense, Ministry of Emergency Situations, Ministry of Internal Affairs, and FSB, as well as individual scholars and scientists. The workshop addressed nuclear, chemical, biological, cyber, and other forms of terrorism, topics that had not been widely discussed in open meetings in Russia. Box 4-3 highlights some topics for future U.S.-Russian collaboration suggested at the workshop.

In view of the events of September 11, some three months later, it is clear that the discussions provided a worthwhile introduction to a subject that was growing in importance for many participants from both countries. At the same time, the American participants had opportunities to meet with Russian specialists who had not previously been involved in U.S.-Russian activities, and these interactions opened interesting possibilities for future cooperation. The proceedings of the workshop—published in both English and Russian—have been in considerable demand in the two countries (NRC, 2002b). Even two years after the workshop, the RAS was searching for any remaining copies to respond to requests for the proceedings.

In December 2001 a smaller workshop on high-impact terrorism was held in Washington. Its primary purpose was to chart a course for future interacademy cooperation in the field. The charter that was developed for this cooperation appears in Appendix C.

Building on this momentum, the academies agreed to give high priority to cooperation on countering terrorism as reflect in the agreement that

BOX 4-3 Topics for U.S.-Russian Collaboration in Counterterrorism

The following suggestions for future cooperation were presented at the interacademy workshop on counterterrorism, held in Moscow in June 2001:

- studies of the many dimensions of information security, including clarification of the importance and scope of national strategies to improve protection of critical networks and identification of areas where international cooperation should be strengthened
- assessments of the types of potential terrorist threats directed at facilities that produce or store dangerous industrial chemicals
- development of methodologies for evaluating engineering and other security enhancements that will reduce the vulnerability of a broad range of industrial facilities (e.g., nuclear power plants, gas pipelines, airports, metallurgical plants)
- consultations of experts on the technical aspects of both marking and tagging of explosives, including recordkeeping requirements for taggants and the associated costs
- development of new concepts for more cost-effective destruction of poorly secured chemical weapons stockpiles in Russia
- investigations of the feasibility of terrorist groups assembling radiological weapons and methods for preventing and detecting such activities
- consideration of the technical details of discriminating between natural outbreaks of diseases and the acts of bioterrorists as well as consideration of the preparations for dealing with the consequences of a bioterrorism attack
- studies of methods for preventing and achieving early detection of animal diseases and for determining the cause of disease outbreaks
- studies of the role of the mass media in terrorism situations and in shaping public attitudes toward terrorism
- joint activities aimed at adapting to the Russian environment the U.S. experience in training specialists to deal with terrorism, in developing organizational mechanisms for coordinating activities of many organizations in preventing and responding to terrorist attacks, and in using forensic techniques to assist in the search for the instigators of terrorist acts.

Source: NRC (2002b: 268–269).

appears in Appendix F. In March 2003 the academies organized another set of meetings and consultations in Moscow. One-day workshops were held on the topics of urban terrorism and cyberterrorism. After the workshops, senior officials and experts from the two academies reviewed interacademy programs already under way, surveyed the landscape to discern recent trends in terrorism, and identified several activities for future attention by the academies. Again, the attendance and interest of Russian specialists from many organizations were impressive, reflecting the higher priority being given to the topic. Among the areas of great concern to the Russians, as indicated in documents presented at the workshop, were terrorism related to transportation systems, cybersecurity, bioterrorism, civil defense responses to terrorist attacks, and the international legal framework for dealing with terrorism. Individual Russian experts also singled out for attention radiological terrorism, electromagnetic terrorism, and protection of chemical storage and production facilities, particularly chlorine-related facilities.[8] Activities suggested for further development are listed in Box 4-4

Directly related to the workshop was the development of an interacademy project on radiological terrorism, emphasizing the threats posed by inadequately controlled sources of radiation used in industry, health, and research organizations, and by radioactive material, such as discarded radioactive wastes packed around high-explosive bombs—commonly called "dirty bombs." The purpose of the project is to recommend priorities for U.S.-Russian efforts to address these problems, not only in the United States and Russia but throughout the world. These priorities will be considered by the DOE as it develops its plans to support efforts to reduce the threats of radiological terrorism.

DISPOSITION OF SPENT NUCLEAR FUEL AND HIGH-LEVEL NUCLEAR WASTE

Although experts designated by the academies first met in 1996 to discuss interacademy efforts to address problems in the field of high-level nuclear waste disposal, it was not until 2002 that the NRC succeeded in persuading the DOE to support such an effort. In an unusual approach, the NRC established a committee composed of five Americans and five Russians to analyze the problems attendant in the two countries to the disposition of spent nuclear fuel and high-level nuclear waste, with special attention to the

---

[8] The proceedings of this workshop will be published in 2003.

BOX 4-4 Potential Areas for U.S.-Russian Cooperation in Counterterrorism

After the March 2003 workshop in Moscow on counterterrorism, the following interacademy working groups were established:

- *urban terrorism:* vulnerability and means of protection of emergency operations centers; modeling the vulnerability of city infrastructures; responses to different warning levels; protection of chemical facilities
- *radiological terrorism:* a road map for intergovernmental cooperation in this field; pilot test area in Moscow for identifying and tracking the presence of nuclear material
- *bioterrorism:* distinguishing between natural and man-induced outbreaks of diseases; epidemiological expeditions to understand distribution of dangerous pathogens; improved methods for rapid detection and analysis of pathogens
- *cyberterrorism:* strategy for developing a robust intellectual community in information security; cybersecurity in the banking sector
- *roots of terrorism:* ethnic relations and multiethnic violence; demographic trends and the spread of terrorism
- *role of the nongovernmental sector:* structuring government-private sector relations to fight terrorism.

Source: Adapted from NRC (2003c).

---

end-points for such disposition. Despite language, administrative, and indeed policy problems in addressing such an important topic through this mechanism, the committee issued a report that received very favorable reviews from experts in the field (NRC, 2003b). Some suggested areas for U.S.-Russian cooperation include:

- assuring the current and future availability of the expert scientists, engineers, and technicians needed to work on spent nuclear fuel (SNF) and high-level waste (HLW) management
- protecting materials useful in nuclear and radiological weapons
- consolidating nuclear materials in a few reliably protected sites
- handling the legacy wastes from nuclear weapons production
- transporting SNF

- developing standard, highly durable waste forms for immobilization of different types of HLW
- developing methods and techniques for extraction of HLW that has been stored in tanks for decades
- developing unified approaches to selection of geological media and sites for the long-term storage and disposal of HLW and SNF
- conducting research and development on methods of processing SNF that produce much less radioactive waste than the PUREX process (NRC, 2003b: 11–12).

In a related activity, in May 2003 the academies organized a workshop in Moscow on the technical aspects of the international spent fuel storage facility that Russia plans to establish. Although the very concept of spent fuel being imported by Russia, even for a limited period of time, has been surrounded with political controversy in Russia, the workshop provided expert views on some of the most important aspects of the design and operation of a facility that meets international standards. The United States controls the movement of about 85 percent of the fuel being considered by the Russians (U.S.-origin fuel) for import. But the United States has tied its approval of such shipments to Russian concessions in its nuclear dealings with Iran. However, this political dimension did not detract significantly from the technical discussions. Several of the most interesting presentations dealt with the following topics:

- legal and technical aspects of importing, transporting, and storing spent nuclear fuel in Russia
- repository site selection: environmental, geological, geochemical, demographic, and access issues
- existing and required physical infrastructure at the candidate sites
- shipping and transportation within Russia and interface with international shipping requirements
- reprocessing technologies: experience with existing technologies and research and development programs
- reduction and disposal of high-level waste, including approaches to transmutation and to geological repositories.

According to Russian colleagues, this modest workshop activity immediately added momentum to the efforts of both political leaders in the parliament and leaders of the scientific community to establish an "objective" mechanism for guiding the development of the international storage facility. In July 2003 President Vladimir Putin established a special commission, chaired by Nobel Laureate Zhores Alferov, to this end, and some of the Russian participants in the workshop were named to the commission. Although the idea of such a commission was not new, the workshop apparently was a significant factor in its formal establishment.

INSIGHTS FROM INTERACADEMY CONSIDERATION OF SECURITY ISSUES

Security-oriented projects will succeed only if the governments of the two countries are prepared to identify and make available the relevant unclassified information and to facilitate access to government experts working on the issues under consideration. Thus far, the academies of the two countries have done a good job in working with the concerned government organizations of both countries. But even under the best of circumstances, the academy efforts will have only a limited impact given the large government efforts—including classified ones—being devoted to these security issues. Academy projects can be useful, however, in stimulating the governments to focus on certain issues, in setting forth approaches that might seem "outside the box" to government officials, and in providing support for policies deemed to be sound. The challenge is not to simply tread ground that has already been thoroughly plowed in intergovernmental consultations. There should be a reasonable likelihood that nongovernmental discussions will help to overcome points of contention.

At the same time, interacademy efforts can often make a significant difference in the long run by documenting the academies' conclusions and recommendations in publicly available reports. Sometimes, and particularly in the long term, these reports provide a basis for both government officials and the public to debate issues in a more informed manner than might otherwise be possible. And as government officials change assignments, the reports can be helpful in the education of their replacements. Particularly receptive audiences for past reports have been the members and staffs of the U.S. Congress and the Russian Duma. In summary, whenever appropriate, greater attention should be given to preparing and disseminating persuasive documentation, in both Russian and English, to constituencies of influence. There is no doubt that a bound National Academies Press publication and its translated companion report attract far more attention than unbound letter reports, academy statements, or unbound manuscripts.

A final guideline is that the academies should be highly selective in choosing security-related topics for their attention through interacademy channels. The difficulties encountered in carrying out interacademy activities involving sensitive topics are manifold, and it is very easy to overload the capacities of the academies in the security field, resulting in ineffectiveness. Success should not be measured by the number of activities that are under way but rather by the quality of the products arising from the activities.

In a broader sense, security considerations have always surrounded U.S.-Russian scientific relations, initially manifested in decisions to grant or deny visas that might provide access to sensitive technologies. Then, in the 1990s, the U.S. government became much more concerned about proliferation and backed up its concerns with funds to develop more secure systems to contain sensitive materials and information in Russia. At the turn of the twenty-first century, security concerns extended to joint efforts in counterterrorism. And now there is new appreciation of the security implications of the health and stability of all elements of society. The academies in the two countries are beginning to respond to this ever-expanding security agenda with programs aimed at public health and ethnic relations as well as nuclear, biological, and terrorism issues.

5

Supporting Innovation:
From Basic Research to Payment for Sales

*Every business participates in technological change as an originator,
user, or victim of technological invention and innovation.*
National Academy of Engineering, 1992

In 1999 the National Research Council assembled a panel of American scientists, economists, and experts on Russia to consider the future of U.S.-Russian interacademy cooperation. They immediately questioned whether the United States in general and the NRC in particular should waste time and effort addressing problems in Russia in view of the fact that the indigenous Russian technology was unlikely to benefit the United States or even contribute significantly to economic development in Russia. According to the panelists, globalization of the process of development and use of modern technology was taking place in Europe and Asia. The NRC should therefore concentrate its efforts in those regions. They went on to argue that Russia will not reemerge as an industrial power anytime soon, but will remain primarily a source of natural resources and high-quality cheap technical labor that can be exploited from afar. They concluded, nevertheless, that the "intangible" benefits of interacademy cooperation were manifold and included strengthening political relations between the two countries. Therefore, cooperation in civilian-oriented areas should continue and indeed expand.[1]

---

[1]This meeting of experts was held at the National Research Council in March 1999.

In addition to the security reasons rooted in many aspects of science and technology for U.S. involvement in Russia (see Chapter 4), several responses can be made to the limited vision of near-term technology initially set forth by this group of experts. These responses include the following:

• The scientific base of Russia may be in a fractured state, and half of the vast repositories of industrial equipment may be obsolete, but many examples of significant Russian contributions to international science and to the development of the products of multinational companies exist, even during the dismal 1990s. Although Russian technological prowess has become badly tarnished, a few well-honed patches shine through. The Russian leadership is determined to show that technology born and used in Russia can again become its engine of growth without the need for Western assistance from the international development banks or from foreign aid agencies. In the view of a somewhat overly optimistic but headstrong Russian government, there are many hopeful signs that the economy is beginning to move from one that is simply a source of natural resources and cheap technical labor to one that is knowledge-based as well. Determination, when followed by demonstrated commitment, will be a good first step toward success.[2]

• Russia occupies one-seventh of the earth's land surface, and many developments in Russia affect the United States. The release of environmental pollutants in the Arctic and in the world's oceans, the cutting of forests serving as carbon sinks that help control greenhouse gases, and the spread of HIV/AIDS and other infectious diseases are all global issues that cannot be ignored. Also, such a large landmass will remain astride many international communications and transportation routes that rest on properly functioning modern technology, particularly at the nodes. And trade with many bordering countries, while currently limited, will have a growing influence on world markets.

• It is better for U.S. organizations to be actively engaged in investment activities and cooperative programs in a variety of fields than to sit on the sidelines, constantly guessing the future of the large, untapped Russian market, reacting to the next Russian technological surprise—whether it be a startling achievement or a technological catastrophe—or assessing the emergence of new science and economic partnerships with countries with

---

[2]For an assessment of the state of industrial technology in Russia at the end of the 1990s, see McKinsey Global Institute (1999).

questionable motivations (e.g., Syria, China, Iran). Technology is driving worldwide developments, and it makes sense for the U.S. government to attempt to influence the emergence and spread of Russian technologies in directions that promote U.S. interests.

During the 1990s, the United States and many European governments agreed with this analysis and supported a large number of science and technology cooperative programs with Russia. They also aggressively promoted Western investments in Russian industrial development that they thought would turn a profit.[3] They may have dismissed as unrealistic bravado some Russian predictions about the revival of Russian industry and agriculture, and they may have seen many cooperative projects fail to produce the promised results, but Western technology hunters, encouraged by their governments, have steadfastly observed research and production facilities, watching and waiting for a technology revival while keeping a tight rein on their pocketbooks.[4]

In the late 1990s, both Western government and private sector organizations began to move toward strengthening ties with individual Russian specialists, while minimizing the direct involvement in cooperative projects of the leaders, accountants, and midlevel managers of the Russian institutes or companies where the specialists were employed. In the view of the Westerners, these "overhead" people would only reduce the amount of funds available for real work. Salary payments were therefore increasingly made directly to the Russian specialists and not through their organizations. The specialists were provided with trips to the United States and Western Europe, and they were even given credit cards and health insurance by some Western companies.

In some of its programs, the NRC adopted the approach of supporting individual scientists while seeking only the general approval of the management of their organizations. Indeed, almost all Western organizations, including the NRC, assumed that the Russian institutions serving as the base of operations for the highly talented Russian specialists would somehow find the funds needed to continue their operations. Overhead and maintenance of the physical infrastructures of organizations were exclusively Russian problems, contended most Western collaborators. Even when telephones

---

[3]See, for example, OECD (1994).
[4]For a discussion of innovation problems that must be addressed if this revival is to occur, see OECD (2001).

were disconnected and electricity was turned off at Russian facilities no longer able to pay for these services, these Western partners remained firm in their insistence that the problems had to be solved by the institutions themselves.

INDIVIDUAL EXCHANGE PROGRAMS

In recent decades, the NRC conducted five exchange programs between individual specialists from the United States and Russia. Three of these programs, which seemed important at the time, were short-lived. The funding bases were simply too weak to sustain the programs, because the priorities of the Departments of State and Energy and U.S. Agency for International Development, which were providing the financial support, were in a state of constant change. Moreover, the applicant pools for these programs were not deep. Finally, there were concerns that contacts established during the programs might pave the way for emigration to the United States. A fourth program of individual visits by biology researchers, which began in 2001, is still in its early stages. The fifth program, representing a modification of the interacademy scientific exchange program established in 1959, continues. A change in its objectives, an adjustment in its manner of implementation, and the continued interest of the National Science Foundation in providing support are among the reasons the program has continued.

Three Short-Lived Programs and One New Start

Beginning in the late 1980s, the U.S. Congress decided that the pool of American specialists with substantial knowledge of developments in the former Soviet Union should be expanded, and it has provided the Department of State with funds ever since to achieve this objective. Earlier chapters described several young investigator programs initiated by the NRC under this program. These activities involved sending groups of 6–10 specialists to the former Soviet Union and Eastern Europe to become familiar with developments in selected areas and to establish contacts that could lead to sustained collaboration. A variation of this initial approach was adopted by the NRC in the 1990s whereby individual American specialists interested in different policy issues—including policies with scientific dimensions in the environmental, economic, security, and other areas—traveled to the former Soviet Union, including Russia, and Eastern Europe, to pursue self-designed projects. About 20 scholars went to

Russia for two to six weeks over a period of several years. Almost all reported positive results in terms of publications and valuable new contacts from their interactions.

A second program brought 13 Russian researchers in scientific areas directly linked to nuclear waste management to U.S. universities for stays of up to one year. The NRC selected these participants on the basis of their knowledge of technologies that the Department of Energy might consider using in the United States. Although the pool of potential Russian participants with skills of direct relevance to the solution of U.S. waste problems seemed to be large, the number who could leave their positions in Russia for long periods was limited. It was simply difficult for the NRC to compete for their time with the more robust long-term programs of nuclear engagement being supported by the International Atomic Energy Agency (IAEA), the European Union, and Russia itself. Moreover, American scientists were often reluctant to serve as hosts for long periods without additional salary funds for themselves.

Early in the 1990s, USAID proposed that the NRC establish an exchange program for applied scientists from the former Soviet Union who would work in the United States on problems related to economic and social development. After USAID rejected an alternate NRC proposal that called for building centers of excellence in Russia and other former Soviet republics, the NRC acquiesced to USAID's interest in bringing the specialists to the United States. The program was named CAST (Cooperation in Applied Science and Technology) to help distinguish it from the COBASE program (Cooperation in Basic Science and Engineering) discussed later in this chapter.

The response of Russian applicants to the opportunity to spend time at U.S. universities reflected an intense interest among a few outstanding scientists in Russia both in staying abreast of U.S. achievements and in seeking alternatives to the decaying conditions in Russian laboratories. Among the applicants were many former Soviet weapons scientists. An internal NRC review of the five-year program gave high marks to most of the 150 exchange visitors—the participants were well qualified; 50 percent were under the age of 40; and the visits helped the host institutions to leverage other funds to expand or continue the collaboration. Over the longer term, however, reports of few opportunities to use newly acquired skills in Russia and of plans by Russian participants to emigrate began to increase. The program was terminated as USAID shifted its funding priorities in the hope that at least a few of the participants who had been supported would become important pillars of a revived research base in their home country (NRC, 1996b: 1, 1997d: 1).

A fourth program of individual visits in the field of biological research began in 2001. It is too early to assess whether the approach is sound and the impacts significant, however. Specifically, at the request of the Department of Defense, the NRC has supported a dozen American biology researchers to undertake two-week visits to Russia to determine whether any activities under way on dangerous pathogens at former biological defense laboratories would be of interest to the Americans as topics for long-term collaboration. About half of the Americans have made useful contacts in Russia and have considered next steps for serious collaboration. However, such collaboration becomes complicated because of the sensitivity of the topics and the overlaps with other U.S. government programs at the facilities. Nevertheless, dangerous pathogens are an important subject, and discussions among government officials, host laboratories in Russia, and the American investigators are under way to chart a path for this type of collaboration.

The Fifth Program: COBASE Continues Support of Basic Science

The fifth program—Cooperation in Basic Science and Engineering (COBASE), supported by the National Science Foundation—continues the exchange program born in Soviet times, but in a significantly modified form. This travel grant program is administered on a regional basis to include specialists from most of the states of the former Soviet Union and from Eastern Europe teaming with American scientists. Although only 10–15 U.S.-Russian teams are selected in this competition each year, the quality of the participants is quite high, and the program has considerable value, symbolizing the importance of U.S.-Russian exchanges in the basic sciences. In the 1990s, two significant changes were made in the program's approach: (1) a new emphasis on supporting projects that have a likelihood of leading to more ambitious follow-on exchanges supported directly by the NSF, and (2) limitations of eight weeks abroad for each exchange, although the eight-week period can be broken into two separate trips in either direction to better accommodate other commitments of the participants (the approach before these modifications called for a single trip and allowed for visits of up to six months).

Seeking to gauge the impact of the program, the NRC staff routinely sends questionnaires to the American participants one year after their ex-

TABLE 5-1 Results of Surveys of American Participants in COBASE Program (percent)

	Survey Year		
	FY 1998	FY 1999	FY 2000
Partners still in contact	95	94	95
American had publication or presentation based on program	69	67	58
American applied for follow-on grant	52	67	56
American received follow-on grant	24	33	28

Note: Surveys were conducted one year after completion of individual programs. Although the data cover exchanges involving a number of countries, exchanges involving only Russian scientists had about the same results.
Source: COBASE program data.

changes. The questionnaires ask, among other things: Has contact between the American and Russian specialists been maintained after completion of the exchange? Has the U.S.-Russian team applied for and received funding from other sources to continue the collaboration? Have the members of the team published papers, made conference presentations, or taken other steps to enhance their careers or contribute to the scientific community as a result of experiences during the exchange? The answers vary from year to year, but in general the lasting impact of many exchanges is impressive (see Table 5-1).

As for trends in the program, the number of female applicants has increased only slightly in recent years, and they compete well for the available grants. Beginning investigators receive preferential treatment in the review process, but the number remains small. The limited duration of the visits of Russians to the United States has reduced the likelihood that the exchanges will encourage emigration, particularly because families almost never accompany short-term exchange visitors (see, for example, NRC [1998a: 17–18]).

The results from individual exchanges such as the ones just described usually need time to materialize, and they are manifested in various ways such as joint publications and follow-on visits of other scientists or students. Occasionally, however, the results are evident immediately. Whatever the case, evaluations and measurements of success should not be based on rigid standards.[5]

---

[5]For an indication of the scientific results of COBASE grantees, see www7.nationalacademies.org/dsc/COBASE_Current_Grantees.html.

COMMERCIALIZATION OF TECHNOLOGY

Many American companies now use Russian technologies, often with some form of U.S. government support. This support might take the form of government financing of feasibility studies, government loans for exports of U.S. equipment, government guarantees that foreign investments will not be nationalized, or diplomatic intervention when a commercial deal encounters bureaucratic resistance in Moscow. In the early 1990s, U.S. companies followed several investment routes in Russia, including organizing joint ventures, acquiring Russian firms, or establishing subsidiaries. Now the approaches used by most companies emphasize short-term contracts with Russian enterprises and specialists that minimize the U.S. funds at risk.

In the research arena, Western businesses have adopted a variety of approaches. For example, the Boeing research center in Moscow provides facilities for researchers who remain employees of several Russian institutes. The Corning research center in St. Petersburg hires Russian scientists as Corning employees. In Moscow, the Schlumberger research center combines these two approaches: it funds scientists who remain affiliated with their Russian institutions and employs Russians who work at the center.

The U.S. government has established programs within the Departments of Commerce, Energy, and State to encourage linkages between Russian researchers and U.S. firms. Although the importance of the training aspects of these programs cannot be denied, establishment of commercially viable linkages is a difficult challenge. Often the programs emphasize "technology push" (sometimes at Russian insistence), whereby Russians with technologies already in hand search for Western customers. Less often the programs emphasize "market pull," whereby firms set forth their requirements and the Russian researchers respond to these needs.

Since 1996 the NRC has been addressing the commercialization issue in a small but focused way. The NRC effort gives priority to improved linkages between Russian researchers and Russian firms, with the firms increasingly outsourcing research projects. The objective is not simply to commercialize technologies, but also to provide new job opportunities in the Russian manufacturing sector—opportunities produced by developing upgraded technologies locally for use in Russia. Many Russian researchers want to market their technologies abroad, but they should keep in mind the difficulties frequently encountered by Russian organizations that do not have stable customer bases in Russia: they are trying to operate abroad where many markets are fickle about suppliers and demand fluctuates.

In the early 1990s, the National Academy of Engineering and the Russian Academy of Engineering (RAE), with the participation of the Russian Academy of Sciences, organized a workshop in Russia on the management of technology. The RAE assembled a group of industrial managers who, with considerable nostalgia, recalled earlier Soviet experience with systematically pushing technologies from research institutes to state enterprises without the constraints of competition and the need for careful economic analysis. The American participants emphasized the new responsibility of individual entrepreneurs to find paying customers or go out of business. Even though all participants agreed on the need for a new brand of management training in Russia, few ideas emerged on how technology development could thrive in the depressed economy of Russia. Thus, in the absence of new suggestions there was little enthusiasm for a follow-on activity (Kershenbaum, 1996).

Several years later, the NRC and the RAS initiated a series of consultations and workshops revolving around the activities of small innovative firms in the two countries. They addressed problems facing Russian entrepreneurs such as changing tax policies, lack of sources of investment capital, and excessive licensing and inspection requirements. Russian counterparts had the opportunity to visit a variety of technology incubators and technoparks in North Carolina and the federal technology transfer center in Wheeling, West Virginia. American specialists consulted with researchers and entrepreneurs in Moscow, Zelenograd, St. Petersburg, Samara, and Obninsk.

In all of these interactions, the interest of potential industrial customers in financing research and development activities was a central issue. Russian colleagues were impressed by the level of financial support provided by the U.S. government, through the Small Business Innovation Research program and other mechanisms, and by state agencies to small firms and research centers. In Russia, government support is often simply a token gesture. American participants were taken aback by the reluctance of many Russian entrepreneurs to engage in discussions of market needs with potential industrial customers before they invested their meager funds into technology development that might or might not find a market niche. At the same time, the Americans noted a generational gap between the old-style Soviet approach of many senior research managers, who felt marketing was someone else's responsibility, and young entrepreneurs, who were willing to risk their own money, but only if they found acquaintances or at least contacts who seemed to be in a position to become paying customers for new technologies that could reduce customer costs quickly (NRC, 1998b, 2002c).

As for the protection of intellectual property, the NRC and RAS can take considerable credit for the slow evolution of Western-oriented approaches in Russia that stimulate rather than discourage development of new technologies. The core issue has been the ownership of technology developed with the use of Russian government funds. There is little hope of untangling the confusion surrounding technology developed during the Soviet period that was incorporated into new products during the 1990s, but there is increasing acceptance of the concept that, for entirely new technologies, the government should give the patent rights at no cost to the institutions where the inventors are employed and that the inventors should receive a fair share of profits from the inventions that take hold. Indeed, as the exposure of key Russian government officials to U.S. approaches sharpened the debate in Russia on the eve of Russia's entry into the World Trade Organization, the Russian government adopted some principles of the Bayh-Dole legislation passed by the U.S. Congress (the legislation bestows on performing institutions the patent rights emanating from government-funded research).[6]

The NRC also can take credit for raising the level of awareness of the importance of market pull as a crucial complement to the traditional technology push approach in Russia. Although only a few Russians have fully accepted this idea, the message is slowly spreading that technologies should be developed to respond to real market needs defined by potential customers. Linked to this important concept has been the call by the academies for the establishment in Russia of industrial-university centers and industrial consortia that would strive to improve the process of using local capabilities to respond to technology needs. Interacademy consultations are sometimes cited in debates over these topics in Moscow.

In 2000 the NRC and RAS launched an ambitious program to demonstrate how both modern approaches to research management and electronic networking technologies can strengthen the linkages between researchers of

---

[6]For a discussion of the patent and related issues that have been addressed in the interacademy program, see Schweitzer (2000: Chap. 5). Discussions confirming the importance of the interacademy program were held with the director of the Russian Patent Agency in April 2003. For a discussion of issues about the Bayh-Dole legislation currently being raised in the United States (e.g., restriction of dissemination of academic research results, diverting faculty from basic to applied research), see Thursby and Thursby (2003).

the RAS and Russian industrial clients—current and future. The idea was to upgrade the capabilities of two innovation centers within the RAS to respond to the interests of Russian companies willing to finance research that improves company profitability. One center, located within the firm Petrocom at the Institute for Control Sciences, was already well along in establishing an impressive customer base. The second center, located at the Institute for the Geology of Minerals (IGEM), was in an embryonic stage.

Petrocom, which is linked financially with the Honeywell company, has many customers within the Russian petrochemical industry who use sophisticated control technologies in optimizing refinery and related activities. Also, there is considerable interest in Russia and abroad in Petrocom's software packages for training refinery operators. The objective in focusing the interacademy effort on this center was, first, to encourage Petrocom to expand its horizons, build on its past successes, and significantly expand its activities to encompass a larger number of clients and a larger cadre of researchers serving these clients, and, second, to provide a successful model that could be emulated in some respects at IGEM and indeed by still more centers in the future. The main focus of the project, however, was on the center at IGEM.

The concept implemented at IGEM quickly attracted interest in both Russia and the United States, and the limited funding available to the NRC for upgrading IGEM electronic networking capabilities was supplemented by funding from Russian, international, and U.S. sources. IGEM followed two tracks as its networking capability was upgraded. First, it strengthened its ties with several paying customers, primarily through direct consultations but also through conferences bringing together government officials, industrialists, and researchers to talk about general concepts for interaction and specific research activities of mutual interest. Second, IGEM scientists began to build digitized databases from the enormous amount of data collected over several decades on the natural resource potential of the country. IGEM wanted to be in a strong position to provide advice to customers on both proposed industrial development strategies and strategies that had not yet occurred to them.

As noted earlier, the results of cooperative efforts may not be evident for years. If this project achieves its original goals, IGEM and its industrial clients should be beneficiaries well into the future. Also, other RAS institutes, having learned about the project, are lining up to emulate the IGEM experience, although each case presents unique requirements and challenges.

In a stunning development in November 2003, the Norilsk Nickel Company committed to provide $30 million annually for 5 years for research support by the institutes of the RAS. The RAS attributes this commitment in significant measure to the interacademy project sited at IGEM. While there had been years of discussion in Russia about the importance of linking researchers and industrialists within the new business environment of the country, the interacademy project demonstrated how this could be done and made a positive impression on the leadership of Norilsk Nickel. (See Appendix H for the National Academies press release on this development.)

Box 5-1 sets forth some of the lessons learned during the IGEM project.

IMPROVING ETHNIC RELATIONS IN RUSSIA

Straddling definitions of security and nonsecurity projects is an interacademy effort to encourage improved approaches to resolving ethnic problems in Russia. In response to a suggestion from the RAS in early 2000, the NAS assembled a small team of social scientists with international experience in addressing ethnic relations to collaborate with several leading Russian specialists concerned about the course of the war in Chechnya. Some participants in the project had worked together in 1993 preparing an NRC report on ethnic conflict that had focused on the general issue rather than Russia in particular (Walker and Stern, 1993). After the American specialists consulted intensively in early 2000 with Russian officials and experts in Moscow and with officials and specialists from the northern Caucasus in Rostov-on-Don, the two academies sponsored a conference in Moscow in December 2000 on ways to prevent eruption of ethnic violence, to bring an end to violence, and to reconstruct societies torn by ethnic-rooted violence. A large number of Russian officials and specialists as well as a few Western experts participated in an open and frank discussion of issues at a time when Russian officials were worried about the outbreaks of violence in various minority enclaves in Russia (or the 10 ethnic hotspots, as the Russian government called them).

Given the high level of interest in the topic among government officials in Washington and Moscow, the academies, with support from the MacArthur Foundation, continued the dialogue and carried out field investigations after the conference. American specialists met with political leaders and ethnic experts from the Volga region in Nizhny Novgorod in October 2001. The successes achieved in preventing ethnic violence in the region

---

BOX 5-1 Lessons Learned in Developing Innovation Centers in Russia

An interacademy project to upgrade capabilities of two innovation centers within the Russian Academy of Sciences provided insights into such centers, which are proliferating throughout the RAS.

- In the natural resources sector, Russian companies are very interested in using the talents of Russian researchers to support industrial needs as long as there are easy routes to tapping into this expertise.
- Direct involvement of the leadership of the Russian Academy of Sciences in the activities of innovation centers lends prestige to the centers that is important in attracting industrial clients.
- Russian industrial representatives prefer to discuss their technical problems with researchers who are expert in fields directly related to industrial interests rather than route their views through middlemen in innovation centers who are not experts.
- A high level of commitment and ingenuity from the managers of innovation centers is essential to overcoming the legal, management, and technical obstacles that inevitably arise.
- Installing or upgrading electronic networking in Russia requires detailed planning by specialists who understand the technical requirements, cost constraints, and realities of implementing projects in Russia.
- Installation of customized networks that depend on the purchase of specialized equipment requires that the network manager be directly involved in all aspects of the procurement process.
- An emphasis on linking Russian researchers with Russian industry is highly desirable, but opportunities for researchers to service Western companies operating in Russia should not be ignored.

Source: Discussions in Moscow, April 2003.

---

were impressive and in sharp contrast to the difficulties in the northern Caucasus because of both the different styles of political leadership and the different histories, particularly since 1991. This initial phase of the program concluded with a conference in December 2001 in Washington that sought opportunities for joint U.S.-Russian research teams to discuss violence, identity, and cross-cultural studies (NRC, 2003a).

During 2002, the program was to focus on two areas—Tatarstan and Chechnya. Brief consultations between NRC specialists and officials and ethnic relations experts in Kazan, Tatarstan, in April 2002 revealed that many of the economic issues with Moscow had been largely resolved and that the major debate of the day was the use of the Tatar language in the schools—a topic that did not fall within the mainstream of NRC expertise. Therefore, the two academies decided to devote their efforts to improving educational opportunities in Chechnya.

An interacademy workshop in Sochi in September 2002, with organizational assistance from the Chechen-led Fund for Humanitarian Assistance to the Chechen Republic, attracted 12 educators from Chechnya, representatives of the United Nations Educational, Scientific, and Cultural Organization (UNESCO) and several international nongovernmental organizations (NGOs), and Russian and American specialists. The views from the field were quite insightful. Although the reports on the physical and psychological conditions surrounding educational efforts were depressing, the commitment of the educators to their profession and to the students was inspiring. At the workshop, the NRC announced a program to support innovative pilot projects to improve educational opportunities in Chechnya.[7]

Chechen educators submitted 15 proposals for pilot projects. Of these, six one-year projects were supported with NRC internal funds at a total cost of $20,000:

- establishing a museum of local folklore that will host a regional folk festival (Grozny Middle School #58)
- using distance learning in mathematics to improve the skills of pre-university students (Chechen State University)
- organizing a student essay contest on methods of settling the conflict in Chechnya and priorities for reconstruction (Chechen State University)
- producing textbooks on Chechen literature for grades 10 and 11 (Ministry of General and Vocational Education of Chechnya and Chechen State University)
- exploring employment opportunities for university students (Chechen State University)

---

[7] Institute of Ethnology and Anthropology (2003); for background material prepared with the support of the NRC, see Institute of Ethnology and Anthropology (2002).

- equipping a sports hall for freestyle wrestling within the physical education curriculum (Grozny Teachers College).

Another direct outcome of the Sochi workshop was the initiation by the Danish Refugee Council, which was represented at the workshop, of activities targeting the Chechen educational system. One project links Danish universities with colleagues in Chechnya.

The overall interacademy program on ethnic relations is scheduled for completion at the end of 2004. A workshop to continue examination of ethnic relations, and in particular approaches to local governance, is scheduled for late 2003, probably focusing on Dagestan, to be followed by wrap-up conferences in Washington and Moscow in late 2004. The results of the pilot projects in Chechnya should be available in early 2004 and reported at the conferences, along with significant observations on lessons learned during the program, to interested officials and specialists.

ROLE OF RUSSIAN UNIVERSITIES

Many Russian leaders believe that the country has lost a generation of well-trained and highly motivated scientists and engineers because of the economic chaos of the 1990s. Although contraction of the pipeline of talented young scientists during the 1990s cannot be easily documented, two critical roles of Russian universities are clear: to continue the Russian tradition of providing graduates with strong preparation in the sciences, and to initiate new business-oriented programs that will improve the likelihood that graduates will prosper financially in an economy in transition. To these ends, most universities need to more effectively link their research and related activities to the activities of the RAS and Russian industry.

Every time leaders of the NAS and RAS have met since 1985, they have discussed the conditions in Russian universities and the importance of joint efforts to revitalize the educational system. But beyond these continuing expressions of concern, the NRC has not been able to design a program targeting Russian universities that would meet the goals of funding organizations. Several other American organizations have launched modest programs at Russian universities (e.g., George Soros's Open Society Institute until 2000, U.S. Civilian Research and Development Foundation until present), but the erosion of science education in Russia presents problems of great magnitude.

GLOBAL ENVIRONMENTAL PROBLEMS

Also high on the agendas of meetings of leaders of the academies has been environmental protection. The global nature of environmental issues is clear, and the RAS plays an important role in Russian government deliberations over environmental policies, particularly since abolishment of the independent Ministry of Ecology. Although the NRC has not been able to develop a sustained program, several ad hoc opportunities for collaborative efforts have arisen.

In June 1995 a group of NRC oil and gas experts traveled to Nizhnevartovsk in northwestern Siberia to learn about oil exploration, production, and transportation issues. The black sea of pollution that spreads over hundreds of square miles of tundra not far from the city and that arises from sloppy production and pipeline practices made an indelible impression on the group as it flew over the watery terrain in a helicopter. On the positive side, the Americans were favorably impressed by the activities at a nature reserve that had been established to demonstrate how oil production, logging, nature preservation, and agricultural activities could coexist in an area of several hundred square miles (NRC, 1995: 16).

Following up on this visit, the NRC and RAS arranged for groups of young investigators from each country to carry out reciprocal visits in two areas. The first group addressed water quality, particularly drinking water safety. This issue had become a major problem, because while Nizhnevartovsk was growing, little attention had been given to the environmental problems in the areas that were feeding the water system. The second group concentrated on sustainable forestry, with the Americans particularly interested in the demonstration project (nature reserve) just described. In terms of continuation of the initial collaboration, the water group reported the most progress; the Russian visitors signed an agreement for long-term cooperation with the University of Massachusetts (NRC, 1996c, 1996d, 1997b: 17, 1997c: 18).

Earlier, in 1993 and 1994, two other reciprocal exchanges of young environmental scientists were held. One group was interested in biodiversity and directed its work toward activities in California and in several Russian nature reserves. The second group addressed Arctic ecology issues. Relevant experience in Alaska and in northwestern areas of Russia was the focal point of this interaction. In later years, several of the participants in each group continued to work with their colleagues with the support of either the NRC individual exchange programs or programs of other organizations (NRC, 1993b: 14–15, 1993f).

The effort to develop a more solid base of interacademy projects began with an interacademy workshop organized in Moscow in November 2000 on the role of environmental NGOs, a topic that appealed both to the environmental community and to the advocates of strengthening the civil society in Russia. By all measures this workshop was a success. The attendance by both Russian government officials and specialists, from Moscow and from outlying regions, was strong; the papers presented by both American and Russian specialists were filled with previously unavailable information; and the free-wheeling discussions gave many Russian environmental activists in attendance an opportunity to have their voices heard by an international audience.

Two other indicators of success emerged as well. First, in preparation for the workshop the RAS established a panel on liaison with NGOs under its standing committee on environmental protection. Today, this panel continues to maintain responsibility for strengthening links between the RAS and key NGOs. Second, the proceedings of the workshop has been in wide demand in Russia and has served as a resource document at various Russian educational institutions. Plans for a follow-on workshop that would focus on a specific region of Siberia, as suggested by Russian colleagues, have languished because of lack of financial sponsorship (NRC, 2001). Meanwhile, the RAS has been using the newly installed electronic networks at the IGEM innovation center to provide improved access to the more than 3,000 reports on ecology that have been recently published by the institutes of the RAS.

COOPERATION ON NONSECURITY ISSUES: LESSONS LEARNED

In 1997 the White House Office of Science and Technology Policy requested that the NRC organize consultations and an open meeting with representatives of U.S. departments and agencies on their cooperative activities with Russian partners in the field of science and technology. More than 100 representatives of the departments and agencies participated, together with about a dozen nongovernmental specialists. Among the lessons learned by the represented groups and that still seem particularly relevant to future interacademy activities are the following:

- Young American scientists are reluctant to take time off from early career development activities to travel to Russia, and thus they need special incentives to encourage them to participate in cooperative programs.
- Reports distributed in the United States about the state of Russian science are overwhelmingly negative, and more publicly available reports on the positive aspects of many research and related activities in Russia are needed.
- Training Russian research and development managers in modern approaches to the effective use of personnel and facilities and to interactions with potential customers should be a priority in cooperative programs.
- The Russian educational system should be strengthened both in training the next generation of science and technology leaders and in supporting Russian government-sponsored industrial technology activities.
- U.S. visa policy requires continuing attention to ensure that it does not inadvertently facilitate an international brain drain by being too lenient in decisions to issue nonimmigrant visas to applicants considering immigration while not being too stringent in the name of national security in issuing visas for international scientific collaboration.
- Concerns about misuse of dual-use technologies for weapons systems are usually exaggerated and should not be allowed to block legitimate commercial deals (NRC, 1998a: 17–18).

Although security-oriented interacademy programs will probably continue to have strong support from both U.S. government departments and private foundations, the likelihood of support for civilian-oriented interacademy programs is less certain. Indeed, the acquisition of funds to support such programs is always a major accomplishment. Meanwhile, the RAS seems more comfortable with civilian-oriented projects, which are in line with its strengths. Yet it is more difficult to promise results of obvious importance to U.S. interests from nonsecurity projects, even though they are less likely to duplicate the efforts of U.S. government agencies than security projects.

Overall, the recovery time for Russian civilian technology prowess will be long and the survivability of many Russian schools of science is uncertain, but the influence of American involvement on both scientific and technology policies and programs can be important.

6

Lessons Learned and the Future of the Interacademy Program

*There is no national science
just as there is no national multiplication table;
what is national is no longer science.*

Anton Chekhov

The leadership of the Russian Academy of Sciences is proud that the RAS thrived in Soviet times and then survived the recent political and economic transitions to remain intact. In fact, it was one of the few Soviet institutions that was not dismembered and completely restructured.[1] This stability has been important to the U.S. National Academies, because they continue to have a responsible and responsive partner in Russia.

Despite the adoption in Russia of new approaches to governance and the economic crisis throughout the country, the RAS rests on three pillars that in many ways closely resemble academy pillars of the Soviet era:

1.   A prestigious, influential, and relatively well-paid membership consisting of academicians and corresponding members. As in the past, most members are selected on the basis of scientific achievements. Efforts by the leadership to achieve election of a significant number of young members have met with only limited success.

---

[1]Comments by the president of the Russian Academy of Sciences at a meeting at the NAS in November 2002.

2.   A large Presidium with an administrative apparatus that controls budgets and senior personnel appointments. In contrast to earlier days, there is only limited central planning of research programs, except when earmarked funds become available for specified topics. These funds are often directed by the leadership to specific laboratories.

3.   Hundreds of research institutes and laboratories. Some are in reasonably good shape; others are in poor condition. Despite greatly reduced budgets, there has been little effort to downsize laboratories beyond not filling positions that become vacant from employee movement to the private sector, emigration, and retirement. The decline in the number of junior and midlevel researchers is of special concern, but a few institutes have found resources to continue to recruit outstanding young researchers (*Rossiskaya Akademiya Nauk 1991–2001*, 2002).

In the years ahead, the extensive property holdings of the RAS will continue to provide considerable rental income for the Presidium and for the institutes. Moreover, the many current and aspiring academicians in influential positions throughout the governmental and nongovernmental sectors will help to ensure the financial viability of the institution. Some members also have strong international scientific reputations and excellent contacts abroad, and they will continue to have seats at the international tables of science. Thus, the RAS will probably continue to weather the depressed economy and play an important role in all aspects of science and technology. Many institutes are nevertheless still in dire economic straits, and they will continue to lose ground as important international science partners.

A clear trend within the RAS institutes during the past decade has been a greater role for applied research and a decline in basic research despite the commitments of the leaderships of both the nation and the RAS to retaining strong fundamental research capabilities. Driven by the need to find commercial sources of financing and by funders' waning interest in basic research that has no economic, environmental, or social payoff in the foreseeable future, most senior officials of the RAS and its institutes agree that even the most brilliant Russian scientists must learn to break bread with paying customers. They realize that technology-oriented entrepreneurs are an important key to a knowledge-based economy that complements the country's historic reliance on exports of natural resources. They also are aware that if the most promising young entrepreneurs are to be sufficiently motivated to take the risks that could lead to business success, these young entrepreneurs must feel secure in meeting their needs for housing and receiving adequate financial rewards for their achievements. On paper, basic research is vital, but in practice applied research is the priority.[2]

Although they are not ignoring the economic downturn, most Russian academy leaders are optimistic about the future of Russian science and technology. They argue that (1) they have survived the worst times; (2) the government understands the problems inhibiting innovation efforts; (3) investors are open to new ideas; and (4) the existing, still considerable science and technology potential can undergird development of new products for both the domestic and foreign marketplaces.[3]

The RAS will continue to be interested in international collaboration in many areas, and especially collaboration that brings financial benefits to the Russian participants. Of special significance, a decade ago many RAS institutes were hesitant to become involved in the security-oriented activities being promoted by foreign colleagues, but such programs are no longer strange to them. Now a surprisingly large number of academy researchers who had little involvement in Soviet defense activities are being recruited to participate in counterterrorism, nonproliferation, and other security-related programs with extensive international dimensions. Also, in both security and nonsecurity areas some institutes have established core programs that are largely supported by foreign organizations in exchange for continuing foreign access to the strong intellectual capabilities of Russian staffs. Other institutes have successfully obtained renewable research and development grants and contracts from abroad. Nevertheless, most institutes are not in a healthy condition.

The changing state of U.S.-Russian government relations during the past decade has had only a minor impact on cooperative programs carried out by the RAS. International security-oriented programs implemented by other Russian organizations such as the Ministry of Atomic Energy are far more sensitive to the political dimensions of bilateral relations. Apparently, Russian president Vladimir Putin places less importance on ensuring parity in the U.S.-Russia nuclear balance, on slowing the march of the North Atlantic Treaty Organization (NATO) toward Russia's borders, and on coun-

---

[2]Some of these views were presented at meetings of leaders of the RAS and National Academies in June 1999, February 2002, and September 2003. However, RAS leaders are reluctant to accept a decline in the emphasis on basic research.

[3]For a snapshot of current research trends and attitudes toward science, see Schweitzer (2001).

tering the projection around the globe of U.S. military power than he does on the information and communications revolutions and the globalization of markets—developments that will directly affect Russia's economic future. In short, Russia has been described as "a center of geo-economics, not geo-strategy, and a pathway for arbitrage and export, not power plays and arms races" (Legvold, 2002–2003). The attendant reorientation of national strategy should call for a greater role in international affairs for science institutions such as the RAS than in the past.

THE VIEW FROM WASHINGTON

All indications are that U.S. policy will continue to emphasize cooperative programs that support U.S. security objectives, promote the oil-related and other commercial interests of American companies, and encourage the evolution of a system of political governance that looks like it was designed, or at least fine-tuned, in the United States. Cooperative science and technology projects financed by the U.S. government will usually be shaped to fit into this three-part agenda. Projects will occasionally address health, environmental protection, small innovative business development, and fundamental science, but clearly efforts to strengthen Russia's institutional capacity to develop and use science and technology effectively in addressing development problems, even if they are of global concern, have not been a U.S. priority. In the months since the February 2003 loss of the *Columbia* space shuttle, continued cooperation in development and operation of the international space station has become a huge question mark.[4]

A major challenge facing the National Academies and other U.S. science-based organizations with international programs is to convince an array of U.S. government departments and agencies that building science and technology capacity in Russia in nonsensitive areas of global concern is an important objective, and that cooperative programs can have tangible benefits for the United States. The Russians have considerable underutilized talent that has contributed significantly to efforts to address, for example, infectious diseases, ocean pollution, global warming, and the search for more efficient energy sources.

---

[4]In 1993 the U.S. and Russian governments signed a 10-year agreement for cooperation in science and technology. It had been scheduled for renewal in 2003. However, U.S. concerns about the possibility that funds flowing to Russia might be subject to Russian taxes and that the United States might be held responsible for accidents or other claims of liability resulting from cooperation delayed such renewal.

The role of the National Academies in mobilizing high-level talent to address difficult issues while promoting U.S. objectives in Russia, and earlier in the Soviet Union, has been widely recognized over several decades. The ongoing financial support of such activities by several U.S. government departments and agencies and private foundations serves as impressive testimony of this recognition. In Russia, the National Academies are well respected because of their high scientific standards at a time when the popularity of foreign organizations is often measured by the size of their financial contributions to Russian organizations and individuals. This latter measurement criterion does not favor the National Academies. They must rely on the quality and results of their activities to keep the welcome mat out in Russia.

INSIGHTS FROM THE INTERACADEMY PROGRAM

Future interacademy programs should take into account experiences of the past decade. Many recent activities have in effect been experiments undertaken in a rapidly changing political and economic environment. A few particularly important lessons learned from recent efforts are described in this section. Most if not all of these lessons are also relevant to other cooperative programs, and the National Academies should certainly share successes and difficulties broadly with interested parties.

Learning from Reviews of Past Cooperative Activities

A review of government-financed U.S.-Russian cooperative efforts is carried out each year by the U.S. coordinator for programs in the former Soviet Union, who has a staff in the Department of State. This review is held in conjunction with the preparation of budget requests by U.S. departments and agencies, so that they can continue or modify intergovernmental programs, which encompass many activities. The review's assessments of impacts are based largely on department and agency self-evaluations. Nevertheless, the annual review provides a framework for considering program initiatives and preserves a historical record of past activities. It also helps the U.S. Congress to assess individual program activities within a larger context (U.S. Department of State, 2003).

A more focused assessment of science and technology cooperation was carried out by the author of this report in 1996–1997 under the sponsorship of the Twentieth Century Fund. The assessment was directed primarily toward programs supported entirely or partially by the U.S. government, including several interacademy projects, although it recognized a few industrial and academic initiatives as well. The report of the assessment sets forth a list of lessons learned from cooperative programs (Schweitzer, 1997). Those that seem particularly relevant to future interacademy activities appear in Box 6-1.

BOX 6-1 A Review of Cooperative Programs: Lessons Learned of Special Relevance to Interacademy Programs

In 1996–1997 the author of this report developed some general principles to guide U.S.-Russian cooperation in science and technology. Among those that will continue to be relevant to interacademy activities are the following:

- Recognize that Russia is different and that many elements of the U.S. model may not be appropriate.
- Replace the concept of technical assistance with the concept of technical cooperation.
- Give priority to the details of implementation of projects.
- Recognize that technical data are considered of great value in Russia and are not given away free of charge, even in cooperative programs.
- Train the real Russian managers—not simply Russians who are seeking training.
- Train in Russia the Americans who manage projects in Russia.
- Do not ignore the support of the Russian research infrastructure in cooperative activities.
- Question the realism of Russian research proposals that may suggest exaggerated benefits.
- Recognize the ability of Russians to develop proposals without prompting from Westerners.
- Find interested audiences before launching demonstration projects.
- Accept the reality and appreciate the impact of the internal brain drain.
- Support both large and small projects.
- Anticipate the ubiquitous tax inspector.

Source: Schweitzer (1997: 98–104).

In addition, as described in Chapter 5, in 1997 the National Academies sponsored a brief review of intergovernmental science and technology programs at the request of the White House Office of Science and Technology Policy. This review suggested steps that could be taken to improve cooperative activities (NRC, 1998a).

Most of the projects supported by the U.S. and Russian academies have been considered successful by financial sponsors and participants. Some have had discernible impacts on the countries' government policies or programs. Yet other projects have encountered difficulties. What are the characteristics of successful projects? Clarity of project goals, timeliness of topic, quality of project leadership, and novelty of approach have often been the precursors of success. Projects can be thwarted, however, by lack of follow-up after interesting contacts have been established, differences of opinion with government officials in both countries on the appropriate roles of the academies, and skepticism of potential funders about the ability of the academies to bring about significant changes in government policies. Clearly, an important lesson learned in preparing this report is that there should be more frequent across-the-board reviews of past interacademy programs to help guide future plans.

Cooperating on Important Topics Not Adequately Addressed in Moscow and Washington

Over the years, a characteristic of many of the most notable interacademy projects has been that they were "ahead of the curve." They explored topics that were of great interest to both governments but that had not benefited from scrutiny by highly qualified, independent experts; they were held in locations not accustomed to receiving foreign visitors; or they involved organizations and individuals who were not regular participants in cooperative programs and who brought fresh perspectives to efforts to address seemingly intractable problems. In the wake of some projects, the governments became interested in sponsoring their own programs in these areas. In the security area, for example, the academies blazed new trails in organizing meetings that attracted biological weapons specialists from closed military research institutes such as the center in Kirov, and in revealing conversion activities in Perm and other cities at defense factories that took "conversion" credit for simply producing samovars and fishing rods. In the civilian arena, the academies concentrated on programs to link Russian researchers and Russian industry, while other international programs in the field of commercialization of technology concentrated on linking Russian researchers with Western investors. A recent pioneering effort has been the establishment of a program of pilot grants for Chechen educators working in Chechnya (see Chapter 5).

Documenting Conclusions from Interacademy Projects

An important characteristic of many interacademy activities has been the participation of specialists from both countries who have had close ties with officials of their governments, thus easing the flow of observations and suggestions from interacademy deliberations to government policy circles, at least during and immediately after the projects. In the mid-1980s, Soviet leader Mikhail Gorbachev established a brain trust of academicians, who were also the principal interlocutors for interacademy programs. More recently, President Putin has turned frequently to RAS specialists who are working with the National Academies for advice on problems ranging from the war in Chechnya to the future of the science cities of Russia. As for the American participants, many have served in senior government positions, some are consultants to government agencies, and others are often invited to participate in policy deliberations in Washington.

This participation in interacademy projects by well-connected specialists is important in ensuring that projects are realistic, that they do not inadvertently duplicate the work of the governments, and that they frame conclusions in a manner that is easily understandable to government officials. However, there is also a down side to excessive reliance on former government officials. Although they are accustomed to participating in important deliberations, they assume that a large phalanx of staff members is available to carry forward their ideas and provide appropriate documentation. This is seldom the case in interacademy deliberations. The academies are not well equipped to become long-term advocates of policies espoused at meetings, even consensus recommendations, in the absence of special efforts to prepare supporting documentation.

Thus, an important lesson is that documents detailing the conclusions and lessons from interacademy workshops, studies, and operational programs can have a considerable impact for many years, while undocumented U.S.-Russian consultations may be quickly forgotten. This is particularly true in Russia, where "hearsay" information is not widely respected and where the turnover rate of government officials who might participate directly in interacademy meetings is high. In the United States, usually a large number of officials in the executive and legislative branches are interested in the topics considered by the academies, and written documentation is the only way to reach many of them. Thus, published reports in both English and Russian are important aspects of interacademy projects and often remain in high demand until the supply is exhausted. Posting reports on the World Wide Web also should enhance the value of the projects.

Engaging the Leaderships of the Academies in Cooperative Activities

The RAS leadership is now playing a bigger role in interagency deliberations in Moscow on many issues. The RAS president is a member of the Russian Presidium and of the Security Council. He also serves as vice chair of the Council on Science and Technology, which is chaired by President Putin.

In Washington, the leaders of the National Academies participate in many interagency discussions, and the staff is regularly invited to discussions on science and technology cooperation with Russia. An example of the close contact between senior government officials and leaders of the National Academies followed the terrorist attacks of September 11, 2001, on New York City's World Trade Center and the Pentagon in Washington. Senior academy officials and experts assembled from around the country were given access without delay to senior government officials addressing vulnerabilities in the nation's physical infrastructure. Moreover, Secretary of State Colin Powell has met with academy leaders, and he delivered a particularly memorable talk on the nation's international agenda vis-à-vis science and technology at the NAS annual meeting in April 2002.

Given these close linkages with governments, the presidents of the NAS, NAE, IOM, and RAS recognize the importance of playing active roles in interacademy programs, and especially programs involving the United States and Russia. They are often in a good position to magnify the impact of interacademy activities in their discussions with government officials and funders. Even if busy with other priority responsibilities, the willingness of the presidents to spend even a few minutes focusing on the details of individual projects can often have a significant payoff in promoting an interacademy effort.

Encouraging Russian "Buy-in" for Concepts Developed Abroad

Since the reemergence of Russia as a state, most interacademy programs have arisen from suggestions from the U.S. side. This bias stems in large part from the fact that the National Academies have taken on the responsibility of raising most of the funds to support the activities. Because most Western funding organizations are interested in activities directed toward transition challenges in Russia, the bulk of the projects are designed either to develop recommendations for implementation in Russia or to support activities that are carried out in Russia.

Therefore, there is a special challenge in ensuring that key RAS and Russian government officials are committed to successful implementation of the projects conceived abroad, including the basic concept of the projects, the details of implementation, and the areas to be addressed by recommendations or by field activities. Such a commitment is often referred to as the "Russian buy-in," a concept widely espoused but often forgotten by U.S. organizations and specialists with their own agendas. Even though informal interacademy discussions usually precede organization of projects, and over the course of projects many dialogues are held about the policy issues or program activities at the heart of the efforts, the true buy-in means that after completion of a project the participants will become active proponents of the project's conclusions. No better example can be cited than the enthusiasm of the leadership of the Russian Patent Agency in carrying forward the ideas on modification of Russian patent laws that were initially tabled at several interacademy meetings in the mid-1990s. Although the significant legislative changes were enacted five years later, the roots of the changes can be clearly traced to the interacademy program.[5]

Sometimes special measures are needed to focus attention on the important issues of the knowledgeable Russians invited to participate in interacademy activities. To this end, the National Academies have frequently commissioned papers, providing small but nevertheless significant honoraria to the Russians who prepare these papers. Also, key Russians are invited to visit the United States as part of project implementation. Both techniques help to ensure that Russian colleagues understand the goals of the projects and have an opportunity to influence the projects' directions and outcomes. The hope, then, is that they will feel a degree of commitment, not only to their own views reflected in the project's report, but also more broadly to the activity in general.

In the United States, the review procedures of the National Academies help to ensure the buy-in of the American participants. Most of the princi-

---

[5]Discussions with the director of the Russian Patent Agency, April 2003. Also see Korchagin and Orlov (2001).

pal participants in projects are the members of NRC committees who are required to approve reports. By attaching their names to reports, they usually embrace at least some of the recommendations as their own.

Emphasizing the Sustainability of Short-Term Projects

Another area of concern is the long-term sustainability of activities undertaken by the academies. Hundreds of U.S.-Russian cooperative projects involving science and technology last for one to three years and then end—often because of limitations on funding—without leaving footprints in the sand. This issue is of crucial importance if the activities are advertised as pilot projects, because no matter how successful the pilot efforts, without follow-up they are considered failures.

Fortunately, sustainability is becoming a high-priority issue for many intergovernmental programs, particularly those designed to enhance security interests. These interests are long term, and even in the security arena, where the academies are currently active, there will come a day when U.S. funding is no longer available.

As for the sustainability of other interacademy projects, the record is spotty, as described in earlier chapters. Given the limited resources available for the continuation of projects and for new starts, the project activities that have been sustained have often had to rely on other organizations in the wake of the academies' efforts. A good example is in the field of high-impact terrorism, where the initial interacademy workshop that attracted many key Russian organizations and several U.S. organizations preceded the terrorist attacks of September 11, 2001, by three months. After September 11, these same Russian organizations were eager to respond to the overtures of other U.S. organizations to engage in cooperative projects in fields that were considered at the workshop. At the same time, they retained their interest in cooperating through interacademy channels.

Adopting Modest Goals for Interacademy Projects

Frequently, false expectations are associated with interacademy projects—particularly when some of the participants are unfamiliar with the traditional role of interacademy cooperation, which, with several exceptions, has emphasized convening specialists rather than initiating operational programs. These false expectations are usually linked to a belief that the academies have easy access to financial resources, and therefore translation of recommendations for additional program activities into action should be relatively simple. Indeed, an important role of the academies is to explore new areas and come up with new concepts for cooperation. But, paradoxically, it is difficult to jump-start new approaches when funding for these approaches is uncertain.

A particularly difficult situation sometimes surrounds interacademy workshops that are addressing complicated issues. All participants know that a single workshop will not resolve the particular problem; rather, it can only start movement toward resolution. Even though the academy organizers will probably emphasize that the goal of the workshop is to stimulate discussions of important topics, to foster new contacts between specialists in the two countries, and to produce a proceedings of the discussions, participants will not be satisfied with such a limited vision and will call for follow-on activities sponsored by the academies.

INTERACADEMY COOPERATION IN THE YEARS AHEAD

In 2002 the academies of the two countries laid out an ambitious agenda for cooperation for the next three years (see Appendix C). Two areas were singled out for special attention in joint statements: nuclear nonproliferation and development of knowledge-based economies (see Appendixes E and G).

The latter topic, a new formulation for the academies, called for activities directed toward integrating higher education with scientific research and industrial development, establishing technology transfer centers, supporting small innovative firms, and all the while protecting the environment. Of special relevance to evolution of a knowledge-based economy are increased understanding of the innovation process (see Appendix I) and trends in scientific manpower (see Appendix J).[6]

Also important in developing interacademy programs are the following technical areas selected by the Russian government ("Basic Principles of the Russian Federation Policy in the Field of Development of Science and Technology for the Period until 2010," approved by President Vladimir Putin, March 30, 2002) for priority in revitalizing the science and technology base:

- information-telecommunication technologies and electronics
- aerospace technologies

---

[6]A good overview of the science and technology potential, organization, legislation, policy, funding, and education in Russia is presented in Gokhberg (1997).

- new materials and chemical technologies
- new transport technologies
- armament, military, and special engineering technologies
- production technologies
- technologies of living systems
- ecology and rational management of nature
- energy-saving technologies.

The interacademy activities agreed to in 2002 stretched the capabilities of the RAS to engage the security-oriented ministries and committees of the country in interacademy projects. Such engagement has involved many exchanges of formal letters and other documents between the RAS and the other organizations. In some cases, the ministries have been reluctant to participate in interacademy activities that they consider to be their responsibilities, although in general the cooperation with the RAS has been excellent. In the nonsecurity areas, the RAS has had less difficulty arranging appropriate contacts, and efforts are being made to consider President Putin's technology priorities in launching new projects.

The number of events in the security-related arena is clearly too high. In 2003, 10 separate events were scheduled in Russia. The RAS simply does not have the staff resources to sustain such an effort in a meaningful fashion. On the U.S. side, there is competition with the activities of government departments, and this large security-related agenda needs to be reduced, at least in terms of the number of events. The most reasonable approach, given the importance of counterterrorism and nonproliferation issues, is to limit the number of events while transforming those that are held into more in-depth activities. Specifically, serious studies of contentious issues should replace single workshops on a topic. Also, a larger proportion of events should take place in the United States to reduce the administrative burden on the RAS.

The focus on security-related issues, while important to the academy leaders in both countries, does not play to the primary strengths of the RAS, which does not have lead responsibilities in Russia in the security area. Therefore, in the years ahead higher priority should be given to nonsecurity interacademy programs. Four themes are suggested:

1. *Innovating for profit.* This theme should build on past experience in developing linkages between RAS researchers and Russian industry, establishing viable high-tech firms in science cities and improving the legal

and financial framework for commercializing technologies in small and medium-size industries. As noted, the academies have agreed to cooperate on the general topic of building knowledge-based economies, which is intimately tied to innovation.[7] And Russia's commitment to high tech innovation was demonstrated in a series of large grants for Russia's most promising technologies in 2003 (Appendix K).

2. *Increasing the interest of the youth in careers in science and engineering.* This theme should extract lessons learned from the success of selected Russian universities in Moscow, St. Petersburg, and the regions that graduate cadres of excellent young scientists who devote their careers to science and technology. Of special importance are opportunities for students to have research experiences and to become convinced that scientific entrepreneurship can have financial payoff. This topic is critical to preventing the loss of yet another generation of technical talent through both internal and external brain drains.

3. *Reducing the threat of infectious diseases.* This theme should extend the interacademy work aimed at redirecting former biological defense scientists to civilian pursuits to include scientists with no previous connections to defense activities. Although the RAS is increasingly engaged in the field of biomedical research, it is important for the U.S. Institute of Medicine and the Russian Academy of Medical Sciences to rebuild their bridges of cooperation through projects that contribute both to public health and to biosecurity.

4. *Promoting democratic approaches to governance in Russia.* The interacademy projects on ethnic relations, environmental NGOs, and the international nuclear spent fuel site have underscored the important role of scientists in clarifying policy options, the consequences of value-laden decisions, and the significance of the evolution of a strong civil society. The RAS is playing an important role in these and related areas, and the ground is fertile for additional activities throughout the regions of Russia directed toward strengthening the civil society infrastructure.

Although the foregoing topics are project-oriented, broader discussions about building the scientific infrastructure both for economic development and

---

[7]An excellent overview of industrial innovation in Russia is presented in Gokhberg and Kunetsova (2001). A good discussion of the role of technology-oriented small and medium-size enterprises and the associated intellectual property issues is included in Watkins, Bossourtrot, and Poznanskaya (2001).

education and determining the role of scientists in a democracy threatened by terrorism will undoubtedly highlight meetings of the presidents of the academies.

In summary, the future of interacademy cooperation is bright. But the National Academies, which for the present must assume the burden of providing most of the funding for travel and related costs, will find the task of raising sufficient funds to support such cooperation difficult, particularly in the nonsecurity area. The National Academies have repeatedly used limited internal funds to support interacademy activities. It is time, however, to put the cooperative nonsecurity programs on a sounder financial footing through more persuasive articulation of the importance of such programs for stability not only in Russia, but also in many peripheral countries where Russian influence is strong.

Epilogue

In the first half of 2003, Russia's economic performance
once again exceeded the most optimistic expectations.

The World Bank, 2003

The upward trend in macroeconomic indicators since the year 2000 suggests that the Russian economy has bottomed out and that slow growth in productivity will in time lead to more and better goods and services (World Bank, 2003). Oil and gas production continue to grow, and foreign debt is declining. Many salaries are higher than 5–10 years ago, and the store shelves in large urban areas bulge with electronic devices, appliances, household items, and recreational equipment produced in Russia and abroad. But in the towns with unprecedented unemployment levels and in the villages with no reliable telephones and unheated schools, there is little optimism for an improved life. Indeed, more than one-third of the nation's population survives on incomes below the poverty level. All the while, declines in the quality of everyday diets and in health services, particularly for children, are contributing to the falling life expectancy. And Russian perceptions of life and death threats to the country are sometimes ignored by policy officials and sometimes are front and center (see Appendix L).[1]

---

[1]Russia includes densely populated industrial regions with well-established scientific and economic infrastructure, production centers in sparsely populated regions, and poorly developed agricultural areas. Living standards vary widely. For a description of poverty levels, see UNDP (2003). Also see WHO (2003).

But despite the hardships, many Russian scientists and engineers maintain a remarkable sense of optimism. They predict a slowing of the internal brain drain that is sapping science and restoration of respect for science and scientists by Russian society, even as they toil in ill-equipped laboratories and outmoded production facilities. The leaders of the Russian Academy of Sciences recognize that the country's science and technology infrastructure is in need of technological resuscitation that will be slow in coming, particularly if they rely only on federal budgets for support of the massive upgrading effort. They understand the importance of paying greater attention to the entire innovation cycle, a process that extends far beyond basic research, beginning with identification of real market needs and concluding only when sales to paying customers are a sustainable reality. Meanwhile, they have begun to attract funds from wealthy Russians to support the research activities of particularly promising young scientific leaders, and several oligarchs support an annual prize of $1 million for science and technology achievements in the field of energy (heralded by the Russian government as the "Nobel Prize for Energy").[2] Even the painful task of systematically downsizing oversized research facilities is beginning as a few biophysics laboratories are singled out as standard bearers for the country (Allakhverdov and Pokrovsky, 2003).

No government that is serious about globalization and the emergence of knowledge-based economies can afford to ignore developments in Russia. For decades, the United States has benefited from the achievements of Russian scientists and engineers, and future benefits are clearly in the offing. It is easy to understand the security arguments for engaging in cooperative programs Russian specialists who have weapons-related experience that might otherwise be directed to parties with hostile intentions. At the same time, the benefits that the United States can derive from bilateral research and development efforts in the civilian arena, while perhaps more difficult to appreciate, also can be profound. The academies in the two countries have an unusual opportunity to demonstrate the importance of cooperation in civilian science and technology. This is not an easy task, but, as we have seen, it is a challenge that can be met through projects that translate concepts into practical applications with near-term payoffs.

The examples of ways in which scientific cooperation has cleared up misconceptions about the activities and intentions of counterparts across

---

[2]Author interviews with the first Russian recipient of the energy prize and with the administrator of prizes contributed by Russian oligarchs for young scientists, June 2003.

the ocean are voluminous. Even a decade after the end of the cold war, scientific travelers in both directions regularly comment that they have improved their understanding of the opportunities and limitations in working together. Interacademy agreements and agreements of other nongovernmental organizations have provided opportunities for track-two diplomacy that provides venues for working together; but it is at the level of the individual scientists and engineers that this concept comes to life. They are in unique positions to address effectively the issues on the frontiers of science and technology that have tremendous political, economic, and security implications, and they can help to push the resolution of the issues in the direction of collaboration—not confrontation.

Meanwhile, for decades both governments have considered the track-two efforts of scientific organizations to be important channels for gathering information that is openly available for the asking, for communication between intellectuals, for gaining insights into what works and what does not work in Russia, and for setting the stage for governmental programs. In a few sensitive areas, the governments maintain a tight leash on scientific interactions, but most of the early fears of the governments that exchanges would be routinely distorted for intelligence or propaganda purposes have disappeared. In a similar change in attitudes, scientists and engineers continue to be aware of the political differences dividing the two countries, but they are increasingly focused on the scientific benefits to be gained from cooperation before applying for their visitor visas. It is precisely this focus on high-quality science and technology that will help to ensure that track-two techno-diplomacy continues to receive broad support as both countries increasingly address the same economic and scientific challenges that face all countries.

Peter the Great, the founder of the Russian Academy of Sciences, wisely predicted in 1724 that "science and education will determine Russia's future." Then several decades later, President Abraham Lincoln who signed the Act of Incorporation establishing the National Academy of Sciences, observed: "I know of nothing so pleasant to the mind as the discovery of anything that is at once new and valuable." With common roots, shared purposes, and joint efforts, the Russian Academy of Sciences and U.S. National Academies have been and should continue to be a force for global peace and prosperity.[3]

---

[3]For descriptions of the origins of the Russian Academy of Sciences and the National Academy of Sciences, see *Statute of the Russian Academy of Sciences, 1724–1999* (1999) and "Founding of the National Academy of Sciences" at www7.nationalacademies.org/archives/nasfounding.html.

APPENDIXES

Appendix A

Highlights of Early U.S.-Soviet Scientific Relations (1725–1957)

1725–1775 — Mikhail Lomonosov, founder of Moscow State University, and Benjamin Franklin gain recognition as the fathers of U.S.-Russian scientific relations.

1775–1800 — Literature is exchanged between Russian and American scientific societies.

— Individual scientists begin to correspond.

— American Philosophical Society in Philadelphia and Academy of Sciences in St. Petersburg elect honorary foreign members.

1800–1860 — First scientific exchange visits are held.

— American scientists travel to Russia to learn about explorations of Siberia and the Arctic Sea.

— Russian mathematicians, naturalists, and linguists attract the attention of American scientists.

— Systematic contacts develop as university networks and specialized scientific research centers emerge.

— Astronomy school is founded in Russia, and Pulkovo Observatory attracts American physicists and astronomers to spend extended periods working in Russia.

1860–1870 — U.S.-Russian ties in astronomy grow.

1865 — U.S. optical firm, Alvin and Company, constructs a large telescope-refractor for the Pulkovo Observatory.

1860–1870 —— American scientists conduct expeditions in northeastern Siberia and the Far East.

1872–1876 —— Russian geographer A. Voevakova visits the United States to research its northern and southern parts.

1876   —— D. I. Mendeleev, a chemist and founder of the periodic table, visits an industrial exhibit in Philadelphia.

1876–1900 —— Frequent reciprocal visits are made by U.S. and Russian scientists.

1890s   —— International Geological Congresses stimulate increased ties between American and Russian geologists.

Pre-1917   —— The original school of physiological research of I. P. Pavlov resulted in many ties in the field of physiology. Many American followers of Pavlov emerge and make numerous visits to Russia.

Post-1917 —— Strained relations reduce regular contacts between Soviet and American scientists.
           —— American scientists assist in recovery from the devastation during the October Revolution.

1922   —— The National Academy of Sciences and the Smithsonian Institution send scientific literature to the Academy of Sciences of the USSR.

Early 1920s —— Ties between American and Soviet societies and individual scientists are renewed, and normal prewar correspondence resumes.
           —— Reciprocal visits are reactivated despite lack of diplomatic relations.
           —— Soviet scientists I. P. Pavlov, V. I. Vernadsky, N. A. Maksimov, N. I. Vavilov, A. F. Ioffe, and P. S. Aleksandrov visit the United States.
           —— American scientists become interested in Soviet developments in the physiological and agrobiological sciences.

1928–1930 —— American scientists conduct a zoological expedition in the Soviet Union.

1932–1933 —— American scientists conduct archaeological excavations in the Soviet Union.

1930s   —— Herman J. Muller, an American geneticist, spends an extended period in the Soviet Union.

Post–WWII —— Substantive contacts come to a complete end.

Late 1940s —— Attempts are made to renew scientific contacts.

        — Ideological conflicts cause cooperation to be short-lived.

Early 1950s — Contacts begin to be restored.

1956   — Many American scientists take part in a conference on high-energy physics in the Soviet Union.

1957   — Turning point is reached in U.S.-Soviet scientific relations.

1956–1957 — Both countries receive about 50 scientists representing various fields.

Source: Information originally provided by Academy of Sciences of the USSR. Adapted from NRC (1990b: 10).

Appendix B

Agreement on the Exchange of Scientists between the National Academy of Sciences of the USA and the Academy of Sciences of the USSR (1959)

In accordance with the Agreement between the United States of America and the Union of Soviet Socialist Republics on Exchanges in the cultural, technical and educational fields dated January 27, 1958 (Section I, Paragraph 2 and Section IX, Paragraphs 1, 2 and 3, of the Agreement) and with the purpose of promoting further scientific cooperation between American and Soviet scientists, the National Academy of Sciences of the USA on the one hand, and the Academy of Sciences of the USSR on the other, hereby conclude the following Agreement:

Exchange of Scientists

Article 1

The National Academy of Sciences of the USA and the Academy of Sciences of the USSR will send approximately 20 persons each from among prominent American and Soviet scientists (at least one-half of whom are to be members of the respective Academies) during 1959–1960 to deliver lectures and conduct seminars on various problems of science and technology as well as for the purpose of studying research work in progress in the USA and the USSR.

Article 2

The National Academy of Sciences of the USA and the Academy of Sciences of the USSR in 1959–1960 will organize, on a reciprocal basis, visits of American and Soviet scientists to acquaint themselves with research conducted in the USA and the USSR (Appendix 1).

Article 3

The National Academy of Sciences of the USA and the Academy of Sciences of the USSR agree to exchange scientists in 1959–1960 for conducting scientific research and for specialization for periods of up to one year (Appendix 2).

Article 4

The exchange of scientists provided for in Articles 2 and 3 of this Agreement may be expanded, reduced, or changed in the course of the fulfillment of the Agreement, by mutual consent between the two Academies.

Article 5

When sending scientists in accordance with Articles 2 and 3 of this Agreement, the sending Academy will notify the receiving Academy at least three months in advance as to the problems of interest to the respective scientists. The sending Academy will also communicate all necessary information concerning the scientists and will indicate the dates desired, the duration of the visit, and the scientific institutions which the scientists would like to visit.

If visiting scientists propose to give lectures, the subjects thereof are to be indicated.

Upon the receiving Academy's acceptance of scientists, the sending Academy will inform the receiving Academy at least 10 days in advance of the date of departure.

Article 6

The National Academy of Sciences of the USA and the Academy of Sciences of the USSR, in addition to the scientific exchanges provided for in Articles 1, 2, and 3 of this Agreement, will invite (on a reciprocal basis) scientists to important congresses, conferences, meetings, and other scientific undertakings of mutual interest.

For this purpose the Academies will exchange twice a year a schedule of such congresses, conferences, etc.

Article 7

The National Academy of Sciences of the USA and the Academy of Sciences of the USSR agree on the desirability of conducting, in the USSR and the USA, joint symposia on current scientific problems in specialized fields.

An organizing committee consisting of representatives of both Academies is to be created for preparing such symposia. A working staff is to be established by the Academy of the country in which the symposium is to be held.

Each Academy shall have the right to publish the proceedings of the symposium in its own language.

Article 8

The National Academy of Sciences of the USA and the Academy of Sciences of the USSR will assist each other on a reciprocal basis in establishing relations between scientific institutions and organizations, archives and libraries, the work of which is related to that of the Academies or is coordinated by them. The Academies will also develop an exchange of scientific publications.

Financial Provisions

Article 9

In all cases, the sending side will defray the travel expenses of its scientists to and from their main destination.

The receiving side will defray travel expenses within its country if these expenses are directly connected with the purpose of the visits provided for in Articles 1, 2, and 3 of this Agreement.

Article 10

The receiving side will provide, free of charge, to the scientists of the other Academy who have arrived on the basis of Articles 1, 2, and 3 of this Agreement, living quarters (hotel accommodations or rooms) and medical aid.

Scientists' salaries (stipends) will be paid by the sending side.

Article 11

Each Academy of Sciences will provide, free of charge, to the scientists of the opposite side who have arrived on the basis of this Agreement opportunities to conduct research in scientific institutions, libraries, and archives.

Article 12

The receiving side will defray the expenses connected with the acquisition of materials, apparatus, literature, photocopies, microfilms, etc., within the program of work agreed upon for the visiting scientists.

Article 13

The expenses of sending scientists to participate in scientific congresses, conferences, and other undertakings conducted in accordance with Article 6 of this Agreement, as a rule, will be defrayed by the sending side, if there is no special agreement to the contrary.

Article 14

Expenses incurred in sending scientists to the joint symposia provided for in Article 7 of this Agreement will be defrayed by the sending Academy.

All expenses connected with preparing and conducting joint symposia will be defrayed by the Academy of the country in which the symposium is held.

Conclusion

Article 15

The duration of this Agreement shall be two years from the date it comes into effect.

Article 16

The Articles of this Agreement may be altered in part by mutual agreement of the Academies.

Article 17

Upon the expiration of this Agreement the National Academy of Sciences of the USA and the Academy of Sciences of the USSR will discuss the question of scientific exchanges for a subsequent period.

Article 18

This Agreement has been signed this 9th day of July, 1959, in two copies each, in the English and Russian languages, the texts in both languages having identical force.

On behalf of the National
Academy of Sciences of the USA

On behalf of the Academy
of Sciences of the USSR

Detlev W. Bronk, President of the
National Academy of Sciences of the
USA

A. Nesmeyanov, Academician
President of the Academy
of Sciences of the USSR

Appendix 1

I. Preferable visits of scientists of the Academy of Sciences of the USSR to study research work conducted in the USA.

1) The study of the latest American spectroscopic apparatus.

1 person for 1 month

2) The study of research work in the field of the theory and practice of automatic control (the theory and practice of automatic regulation, information theory, the theory of relay-action devices, technical facilities in automation and telemechanics).

2 persons for 1 month

3) The study of theoretical and experimental work on durability, plasticity, dynamic problems of plasticity and aeroelasticity, and gas dynamics.

1 person for 1 month

4) The study of work being done in the field of the synthesis of natural and biologically important compounds.

2 persons for 1 month

5) The study of latest researches in the field of microbiology and cytology.

1 person for 1 month

6) The study of research in the field of the biology of antibiotics, vitamins, stimulants of plant growth.

2 persons for 1 month

7) The study of scientific work in the field of photosynthesis.

1 person for 1 month

8) The study of electron-microscope research in different fields of biology.

1 person for 1 month

9) The study of research work in the field of solid state physics and low temperature techniques.

1 person for 1 month

10) The study of research work in the field of radioastronomy.

1 person for 1 month

11) The study of research work in the field of the physical chemistry of polymers.

1 person for 1 month

12) The study of research work in the field of the biochemistry of cancer.

2 persons for 1 month

13) The study of research work in the field of organ and tissue transplantation.

1 person for 1 month

14) The study of research work in the field of epidemiology and the control of influenza.

1 person for 1 month

II. Preferable visits of scientists of the National Academy of Sciences of the USA to study research work conducted in the USSR.

1) Radioastronomy; photoelectric techniques for linear measurements in astronomy.

1 person for 1 month

2) Probability and stochastic processes.

1 person for 1 month

3) Solid state physics and low temperature techniques.

1 person for 1 month

4) Structure and synthesis of nucleic acids; physical chemistry of proteins.

1 person for 1 month

5) Cosmic ray studies.

1 person for 1 month

6) High pressure and high temperature chemistry.

>   2 persons for 1 month

7) Thermodynamics and physical chemistry of igneous rocks.

>   1 person for 1 month

8) Microbiology and cytology.

>   2 persons for 1 month

9) Physiology and biology of the nervous system.

>   1 person for 1 month

10) Biochemistry of cancer.

>   2 persons for 1 month

11) Organ and tissue transplantation.

>   1 person for 1 month

12) Limnology.

>   1 person for 1 month

13) Weather prediction.

>   1 person for 1 month

14) Epidemiology and control of influenza.

>   1 person for 1 month

Appendix 2

I. Preferable visits of scientists of the Academy of Sciences of the USSR to conduct research and specialized scientific work in the USA for periods of up to one year.

1) The study of radiospectroscopy and related fields.

1 person for 5 months

2) The study of new trends and research methods in the fields of steroid compounds, stereochemistry and the chemistry of isoprenoids.

1 person for 6 months

3) The study of research work in the field of high-molecular compounds.

1 person for 5 months

4) The study of experimental work related to the theory of metallurgical processes.

1 person for 5 months

5) The study of work in the field of information theory.

1 person for 6 months

6) Research work to be conducted in biochemistry.

1 person for 5 months

II. Visits of scientists of the National Academy of Sciences of the USA to conduct research and specialized scientific work (in the USSR for periods of up to one year).

1) Solar physics.

1 person for 5 months

2) Non-linear systems and differential equations including applications to mechanical and electric systems.

1 person for 6 months

3) Conditioned reflexes; especially of primates.

    1 person for 5 months

4) Physiology of stress (environmental and emotional).

    1 person for 5 months

5) Celestial mechanics.

    1 person for 5 months

6) Physical chemistry of high polymers.

    1 person for 6 months

Appendix 3

Categories and monthly salaries (stipends) of scientists commissioned to the USSR and to the USA in accordance with Articles 1, 2 and 3 of the Agreement

Scientific position	USSR Monthly salaries of scientists of the USA	USA Monthly salaries of scientists of the USSR
1. Members of Academies of Sciences or leading scientists recommended by the Academies.	5,000 rubles	$500.00
2. Directors of scientific institutions and higher educational establishments and their assistants.	4,000 rubles	$400.00
3. Heads of departments, laboratories, sections and senior scientific workers.	3,500 rubles	$350.00
4. Junior scientific workers.	2,500 rubles	$250.00

Appendix C

Agreement on Cooperation in Science, Engineering, and Health between the U.S. National Academies and the Russian Academy of Sciences (2002)

The National Academy of Sciences, the National Academy of Engineering, and the Institute of Medicine (hereafter referred to as the U.S. National Academies) and the Russian Academy of Sciences, recognizing the many contributions of international cooperation to the achievement of research, economic, and national security goals, will undertake a series of joint activities during 2002, 2003, and 2004. These activities will build on several decades of interacademy cooperation.

Cooperation pursuant to this agreement will be conducted in areas of mutual interest to the academies. The forms of cooperation may include visits of specialists, exchanges of documents including electronic transfers, technical meetings and workshops, seminars and conferences, and joint studies and research projects. The carrying out of these activities is subject to mutual agreement of the academies and to the availability of funds and appropriate personnel.

Activities currently envisaged for 2002 to 2004 are as follows:

(1) The interacademy seminars and related activities on security and arms control issues will continue to receive high priority.

(2) The academies will establish American and Russian committees that will jointly organize seminars, consultations, and related activities concerning the prevention of terrorism and the mitigation of its consequences.

(3) The academies will continue their program of conferences and consultations concerning approaches to reduce ethnic tensions within multiethnic nations and regions of nations and to reduce international problems rooted in ethnic animosities, with particular attention to the relationships of such problems to global terrorism.

(4) The academies will continue to support the study by American and Russian specialists of end points for disposition of nuclear spent fuel and high level radioactive wastes. They will also consider the organization of an interacademy workshop to evaluate the scientific aspects of an international repository in Russia for nuclear spent fuel.

(5) The academies will undertake a study of approaches for maintaining adequate security systems for protecting nuclear material after the termination of U.S.-Russian collaborative programs in this field.

(6) The academies will carry out projects concerning the capabilities of the two countries to develop knowledge-based economies, including the responsiveness of research institutions to industrial needs for new technologies, the role of NGOs in reducing pollution impacts attendant to industrial development, and the importance of linking universities with research institutions and industrial companies.

(7) The Russian Academy of Sciences will continue to facilitate assessments by the U.S. National Academies of project proposals by Russian scientists to collaborate with American scientists on civilian-oriented research on biological pathogens.

(8) The academies will continue to encourage American and Russian specialists in fields such as mathematics, physics, chemistry, biology, geosciences, and engineering to apply for participation in the travel grants program administered by the National Research Council.

(9) The academies will explore opportunities for cooperation on the following topics of broad international interest: scientific and security concerns in Northeast Asia; strengthening of the economies of science cities; and innovative uses of the internet to improve international scientific cooperation.

Other programs of cooperation can be undertaken with the mutual consent of the two Academies.

The financial arrangements for joint activities will be agreed to on a project-by-project basis.

All joint activities are subject to the laws and regulations of the two countries.

The Office of International Affairs of the National Research Council and the Department of Foreign Relations of the Russian Academy of Sciences will serve as the executive agents in the implementation of this agreement.

Each year, representatives of the two Academies will review progress in carrying out activities under this Agreement.

Done in New York, in duplicate this second day of February 2002, in the English and Russian languages, each text being equally authentic.

Bruce Alberts   Yuri Osipov
President, National Academy   President, Russian Academy
of Sciences   of Sciences

Wm. A. Wulf
President, National Academy of Engineering

Kenneth Shine
President, Institute of Medicine

Appendix D

Agreement for Scientific Cooperation between the Institute of Medicine of the USA and the Academy of Medical Sciences of the USSR (1988)

The Institute of Medicine of the USA and the USSR Academy of Medical Sciences, hereinafter called the Parties,

Consonant with the General Agreement between the Government of the United States of America and the Government of the Union of Soviet Socialist Republics on Contacts, Exchanges, and Cooperation in Scientific, Technical, Educational, Cultural, and Other Fields, dated November 21,1985,

Consonant with the Agreement between the Government of the United States of America and the Government of the Union of Soviet Socialist Republics on Cooperation in the Medical Sciences and Health, dated May 23, 1972,

On the basis of the principles and conditions of the Final Act of the 1975 Conference on Security and Cooperation in Europe,

Confirming the mutual interest in cooperation shared by both Parties,

Believing that the rapid progress in science today calls for an enhanced interaction between scientists from both countries based on cooperation by outstanding working scientists in fields in the forefront of science in which the two countries are world leaders,

Recognizing that it is the special responsibility of the Parties to investigate avenues for mutual beneficial scientific communication and cooperation in nonsensitive fields between the scientists of their countries,

While allowing the program to be responsive to the requirement and concerns of the two Parties,

Believing that criteria for selection of fields for cooperation should include the following:

A. The fields are of major scientific importance holding prospects for significant advance;

B. US and Soviet research efforts are in the vanguard and the respective leading scientists would be able to participate actively in cooperation in these fields,

Agree upon the following:

Article I

Cooperation in medical science will be carried out on the basis of programs agreed upon by the Parties in the following forms:

— Mutual scientific research
— Exchange visits of research scientists and delegations,
— Bilateral workshops, seminars, symposia, and lectures,
— Participation of specialists of both countries in international activities hosted by the Parties,
— Exchange of scientific information.

Article II

Consultations of the Presidents and their designated representatives of the Parties shall take place, alternately in the USA and the USSR, at least once a year for the purpose of selecting the fields for cooperation, defining the program, and evaluating progress in implementing the program. Also, there will be discussions of the problems of and opportunities for US-USSR cooperation in areas within the purview of the Parties, and consideration will be given to steps which can be taken by the Parties to contribute to a favorable environment for scientific cooperation.

Article III

1. The quota for the exchange of individual scientists will initially be 12 person-months annually. The quota will be reviewed at the annual Meetings of Representatives. In using the quota, preference will be given to the priority fields identified during the Meetings of Representatives.

2. The Parties will encourage exchange visits by members of the Institute and the Academy. Usually such visits will be for two to four weeks for scientific and public lectures and for scientific consultations. Plans for such exchanges will be defined at the Meetings of Representatives. Within this framework, each Party may nominate its members for exchange visits and may invite members of the other Party.

3. Consideration will be given to exchanges of research scientists for periods of one to twelve months to participate in research activities. The scientists may range from distinguished scientists of international stature to scientists in the early stages of their careers. They should be known by their scientific publications or their participation in scientific meetings. Detailed information about the proposed programs and qualifications of the scientists will be sent by the sending Party to the receiving Party four months prior to the proposed visits.

4. At the Meetings of Representatives, consideration will be given to bilateral workshops on subjects of mutual interest outside the quota. Preference will be given to the priority fields identified during the Meetings of Representatives. Topics and co-chairmen from each Party for such workshops will be determined at that time. The co-chairmen will define the location, develop the agenda, and select appropriate participants for each workshop at least six months prior to the date of the workshop. Generally a workshop will last three to five days and may be followed by visits of one or two weeks to scientific institutions in the host country. The number of participants will be about ten from each Party with additional observers from the host Party. Each Party will have the right to publish a report of the workshop in its own language.

5. Areas for possible cooperative research will be explored during the Meetings of Representatives. Such research programs may involve a variety of activities. They may be carried out under the direct auspices of the Parties or may be recommended to other institutions.

Article IV

The sending Party will provide transportation for visiting scientists to Moscow or Washington, and the receiving Party will cover costs of internal travel and lodging on a reciprocal basis. The appropriate per diem will be determined at the Meetings of Representatives. The sending Party will also provide lodging for dependents accompanying visitors. Emergency medical and dental care will be provided by the receiving Party. The receiving Party will also cover reasonable costs of materials required for fulfillment of agreed research programs.

Article V

The receiving Parties will facilitate the timely issuance of visas to the exchange scientists and their dependents. The receiving Parties will also seek to ensure that the exchange scientists are provided with the necessary documentation for timely departure from the host country, particularly in case of emergency.

Article VI

The Parties may extend or support invitations to scientists of the other Party for special visits, including attendance at national and international conferences and visits to research institutions. The Parties may request the good offices of the other Party to arrange private visits by scientists to scientific institutions in the other country. Financial arrangements for these types of activities will be considered on a case-by-case basis.

This Agreement shall enter into force after being signed by both Parties and shall be valid for a period of five years; at the end of five years, if both Parties agree, it may be extended.

Done in Moscow on January 15, 1988, in two versions, in Russian and in English, both versions authentic.

Samuel O. Thier
President
For the Institute of
Medicine USA

Valentin I. Pokrovsky
President
For the Academy of Medical
Sciences USSR

Excerpt from Memorandum Signed at the Time
of Signature of the Foregoing Agreement

The initial Program will be drawn from among the following themes:

1. Biological and Behavioral Sciences Aspects of Alcoholism and Substance Abuse;

2. Human Immunodeficiency Virus (HIV): Biology, Epidemiology, Prevention, and Treatment;

3. Surveillance and Diagnostic Methods for Polio Suppression; Basic Biology and Development of Poliovirus Vaccine; Molecular Epidemiology of Polio;

4. Health Effects of Ionizing Radiation Releases into the Environment.

Appendix E

Joint Statement by the Presidents of the U.S. National Academies and the Russian Academy of Sciences, February 2, 2002

PREVENTING THE PROLIFERATION OF NUCLEAR WEAPONS AND NUCLEAR MATERIALS

Nuclear weapons or nuclear materials that could be used to develop nuclear weapons or radiological devices must not fall into the hands of terrorists or states with hostile intentions. The United States of America and the Russian Federation, as the nations with the largest nuclear-weapon complexes and the custodians of the largest inventories of nuclear weapons and materials of all types, share a special responsibility for preventing unauthorized access to these weapons and materials.

A decade ago, the governments of the United States and Russia recognized the importance of cooperative efforts to help ensure that nuclear weapons and weapons-grade nuclear materials in the former Soviet Union were adequately protected from theft or diversion. The decision to act came at a time when an economic crisis had seriously reduced the resources available for maintaining security systems and personnel at nuclear facilities. As a result of U.S.-Russian cooperative efforts, thousands of nuclear warheads and hundreds of tons of weapons-grade nuclear material are now better protected. But much remains to be done to place all nuclear weapons and materials under adequate protection.

With clear indications that terrorist organizations are seeking nuclear and radiological weapons, cooperative efforts to deny them this option must be accelerated. These efforts should include plans for the ultimate disposition of the plutonium and highly enriched uranium made surplus by the downsizing of the U.S. and Russian arsenals.

The Academies are encouraged by the recent actions of President Bush and the U.S. Congress to restore funding and a high priority to the joint activities in this domain. They provide the basis for the Russian and American governments to accelerate their cooperative programs to ensure adequate security of all nuclear weapons and weapons-grade material throughout Russia. We urge the two governments to move forward rapidly.

The world has not yet given adequate attention to the dangers of misuse of radioactive sources, spent nuclear fuel, and radioactive waste to make radiological devices. New cooperative activities between the two governments are needed to address these issues—in the United States, in Russia, and throughout the world.

In order to assist their respective governments in all of these efforts, the National Academies and the Russian Academy of Sciences will prepare during the next six months an assessment of the immediate steps that should be taken to upgrade the two governments' collaborative efforts in this domain. Working together, the Academies will develop an agenda for long-term U.S.-Russian cooperation to reduce the risks from nuclear weapons or materials falling into the hands of terrorists or states with hostile intentions. This will include continuing interacademy attention to problems that may arise and how they can be overcome, such as problems associated with access to sensitive facilities. The following interacademy activities related to this assessment and agenda-setting work are already under way or will soon be initiated to provide more detailed insights and recommendations for consideration by the two governments.

- A new project will examine how Russia can develop an effective indigenous, sustainable nuclear materials protection, control, and accounting system. This effort will help the Russian nuclear institutions make the transition for the eventual termination of U.S. financial support of these efforts and it will help the Russian government develop the necessary nuclear legal and regulatory framework and practices.
- An assessment of end points for disposition of high-level nuclear waste is currently under way that pays particular attention to the physical protection of spent fuel and high-level radioactive waste in the United States and Russia.

• A new assessment will examine ways in which U.S.-Russian cooperation on strategies for the ultimate disposition of weapons plutonium and highly enriched uranium can be reinvigorated and enhanced.

• A new project will identify the potential for misuse of radioactive sources available widely throughout industry, medical facilities, and research organizations in the United States, Russia, and other countries.

These joint activities will continue the long-standing cooperation between the Russian Academy of Science and the National Academies in support of their governments' efforts to respond to urgent international security problems.

The Russian Academy of Sciences and the National Academies call on the national academies of sciences of all countries possessing nuclear weapons or using radiological materials to cooperate with them in this most important sphere of national and international security.

Appendix F

Annex 2 to the Agreement on Cooperation in Science, Engineering, and Medicine between the Russian Academy of Sciences and the U.S. National Academies (2002)

RUSSIAN-AMERICAN COOPERATION IN COUNTERTERRORISM

The U.S. National Academies and the Russian Academy of Sciences, recognizing the urgent need for collaborative science and technology-based efforts across the broad spectrum of areas related to prevention, response, and mitigation of terrorism, will undertake a new joint program under the guidance of Russian and American standing committees. The committees will organize joint studies on how to cope effectively with emerging threats and challenges related to terrorism. The program will provide an independent avenue for scientists and specialists to perform studies and analyses, to exchange data and findings, to hold workshops, seminars, and conferences, to train specialists, to educate representatives of the media and other relevant organizations, and to recommend future cooperative programs and projects between appropriate organizations.

Areas of particular interest for this effort may include, but are not necessarily limited to:

- Radiological terrorism, including protection of radioactive sources and wastes;
- Access by terrorists to nuclear materials and technologies and the security of nuclear materials and facilities;
- Bioterrorism against both humans and the food supply, including preventing access by terrorists to dangerous pathogens and application of

new technologies for prevention and detection of terrorist incidents and for responses to them;

- Chemical terrorism, including prevention of access by terrorists to dangerous chemicals and application of new technologies for prevention and detection of terrorist incidents and for responses to them;
- Electromagnetic terrorism and the prevention of damage to electronic equipment sensitive to electromagnetic effects;
- Safety of vulnerable industrial and energy infrastructures and transportation facilities;
- Cyberterrorism, including education and training of specialists;
- Improvement and harmonization of the international and national legal basis for combating terrorism;
- The social, economic, and ethnic roots of terrorism.

In furtherance of the above-mentioned activities, the two committees will commission papers and analyses in specific areas of high priority involving American and Russian specialists with relevant expertise. Likely initial topics will be cyber, radiological, and biological terrorism.

The committees will consist of up to ten members each. The chairs and members of the committees will be approved by the U.S. National Academies and the Russian Academy of Sciences. Their activities will be appropriately coordinated with other interacademy activities and with intergovernmental programs.

Appendix G

Joint Statement by the Presidents of the U.S. National Academies and the Russian Academy of Sciences, February 2, 2002

THE DEVELOPMENT OF KNOWLEDGE-BASED ECONOMIES

The world is moving toward knowledge-based economies. In such economies, science and engineering—linked to marketing, production, and finance—are critical for the operation of efficient and profitable industries. New high-tech capabilities in many fields are essential if industries are to be competitive. Investments in new intellectual capital must increase, and the existing labor force needs new skills both to solve technical problems and to manage technology. Also, the service sector must play an enhanced role to ensure that the economy will serve the people.

The United States of America and the Russian Federation have been the birthplaces of many of the world's most advanced technologies. As all nations move toward knowledge-based economies, the National Academies and the Russian Academy of Sciences will work together to use the scientific end engineering capabilities of our two countries for the improvement of social and economic conditions at home and abroad. The academies intend to concentrate joint efforts during 2002 on the following important aspects of the evolution of knowledge-based economies.

• **Integration of Higher Education with Scientific Research and Industrial Development:** The academies will review approaches in both countries to strengthen the capabilities of higher educational institutions for collaborating with scientific research and industrial organizations, particularly as such linkages enhance the education process and equip graduates to assume responsibilities in such organizations. They will identify successful programs supported by government and by industry, as well as innovative approaches of higher educational institutions that could serve as models for encouraging stronger institutional linkages of this type.

• **Technology Transfer Centers:** The academies will support the development of technology transfer centers within the Russian Academy of Sciences designed to strengthen the ties of industrial enterprises with production facilities in Russia that are in search of improved and new technologies with Russian research organizations that have the relevant research capabilities. The initial emphasis will be on research requirements for enterprises in the fields of natural resource development and petrochemical processing. Also to be explored are new opportunities for the Russian Academy of Sciences to respond to technology needs of enterprises in the fields of biotechnology, metallurgy, chemical engineering, and conversion of military production lines.

• **Small Innovative Firms:** The academies will continue to assess the factors that lead to business success for small high-tech firms, and especially firms located in the science cities of Russia. The need to give greater emphasis to the requirements of industry, and particularly Russian industry (market pull), in addition to the current emphasis on technology push, will be of high priority.

• **Environmental Protection:** The academies will continue efforts to reduce the adverse environmental consequences of industrial development. Of particular interest is the role of environmental non-governmental organizations in contributing to governmental decision-making at the national and local levels.

The academies will share the results of these and related interacademy activities with other interested organizations, including the International Research Council and the Council of Engineering Academies and Technical Societies.

Appendix H

Press Release of the National Academies

Dec. 17, 2003

Cooperation Between U.S. and Russian Science Academies Encourages Russian Investments in Innovative Research

WASHINGTON—A partnership between the U.S. National Academies and the Russian Academy of Sciences (RAS) to strengthen links between Russian researchers and private companies is showing promising results, the U.S. National Academies announced today. An increasing number of Russian companies are now providing tens of millions of dollars annually for applied research that is overseen by RAS. In addition, Russian businessmen are financing hundreds of grants each year for young researchers working in cutting-edge fields.

Last month, for example, the Norilsk Nickel Co., one of Russia's largest companies, made a commitment to provide RAS with $30 million annually for five years to support research on hydrogen energy. Last year the companies Gazprom, Neftegazprovodi, and Neftegastroy, in cooperation with the RAS, set up a number of "innovation centers" at RAS institutes.

"The support of the U.S. National Academies in our efforts to intensify interaction with Russian industry has been a major stimulus in convincing Russian industrial leaders that we are prepared to respond to their needs quickly and authoritatively," said Nikolay Laverov, RAS vice president and the Russian leader of the interacademy partnership.

Albert Westwood, former vice president of Lockheed Martin and chair of the National Academies committee that guided the cooperative effort, emphasized the importance of having scientists work closely with industry. "Effective applied research requires scientists to spend sufficient time in the plants of their industrial clients to understand their actual needs and to identify their specific problems. More often than not, the problems revealed could not have been perceived in the laboratory. Research targeted in this way can be both innovative and cost-effective."

Under the interacademy program, U.S. and Russian specialists have concentrated on two major efforts — the development of a new innovation center at the RAS Institute of Geology and the expansion of an established center at the RAS Institute of Control Sciences. The program aims to strengthen connections between the centers' researchers and existing or potential industrial clients through workshops, consultations, and improved electronic networking capabilities. Both centers have significantly expanded their customer bases since the partnership began. The interacademy effort has brought new attention within Russia to opportunities for improving the technological capabilities of Russian industry, participants said.

For three years the Rutter Foundation in San Francisco has supported the interacademy effort in this area. Additional funding has been provided by the International Sciences and Technology Center in Moscow, the Civilian Research and Development Foundation in Arlington, Va., and the Russian Aluminum Co.

The National Academies provide science, engineering, and medical advice to the federal government under a congressional charter granted to the National Academy of Sciences in 1863.

Appendix I

Innovation in the Russian Federation (2001)

Elements of Innovation (in order of importance)
Acquisition of equipment and machinery
Industrial design
Research and development
Acquisition of software
Personnel training
Market research
Acquisition of technology, including acquisition of rights for patents and
    licenses

Sources of Information for Innovation in Industry
(in order of importance)
Exhibits, fairs, and other advertising events
Consumers
Internal sources of industry
S&T literature
Regulations and standards
Suppliers of equipment, materials, components, and software
Competitors
Industry research institutions
Conferences, workshop, and symposia
Patent Office publications
Academies and universities

Factors Inhibiting Innovation (in order of importance)
Economic factors
Shortage of own funds
Shortage of funds from government
High expenditures
Long payoff period
Excessive perceived risks
Low solvent demand for new products

Production factors
Low innovation potential
Deficiencies in legislation
Low consumer demand
Lack of skilled personnel
Underdeveloped innovation infrastructure
Underdeveloped technology market
Lack of information on market
Lack of information on technology
Uncertainty in timing of innovation
Resistance to innovation

Legal Framework for Promoting Innovation
State Support for Small Enterprises in Russia (Law no. 88-F3, June 14, 1995)
Budget Code of Russia (Law no. 145-F3, July 31, 1998)
Tax Code of Russia (Law no. 146-F3, July 31, 1998)
Status of Science Cities of Russia (Law no. 70-F3, April 7, 1999)
Science and State Science-Technology Policy (Law no. 127-F3, August 23, 1996)
Protection of the Environment (Law no. 7-F3, January 10, 2002)

Source: Centre for Science Research and Statistics (2003) and, for legal framework, Martyushov (2003).

Appendix J

Personnel Trends in the Russian Academy of Sciences

Number of Scientists by Degree

	1991	2001
Doctors of science	7,976	9,307
Candidates of science	29,810	26,415
Other scientists	27,651	17,998

Average Age of Scientists

	1991	2000
Academicians	68.2	70.0
Corresponding members	62.7	64.5
Doctors	55.4	58.3
Candidates	45.5	48.5
Other	38.4	40.2

University Graduates Accepted by Academy Institutes

1991	1994	1997	2001
2,130	1,136	1,894	2,273

Graduate Students at Academy Institutes

	1991	2000
Total	6,141	7,601
Affiliated with industry	3,477	6,012

Pay Levels Compared with Those of Other Sectors (academy is 100)

	1991	2000
Other scientific institutions	110	120
Industry	210	125
Construction	240	160
Overall economy	175	100

Use of Budgetary Resources

Fundamental research	73.8%
Applied research	14.2%
Development	12.0%

Source: Rossiskaya Akademiya Nauk 1991–2001 (2002).

Appendix K
Innovation Projects of National Significance to Be Financed by the Russian Ministry of Industry, Science, and Technology During 2003-2006
(title, performing organization, location, award for total period)

1. **Development of technology and launch of mass production of a new generation of dense and fire-resistant materials for general industrial applications**: Unikhimtek firm, Moscow, 400 million rubles

2. **Development and initiation of production of nanotechnology-related instruments and equipment**: NT-MTD firm, Zelenograd, 400 million rubles

3. **Development of biotechnology and initiation of industrial-scale production of seed stock for high-yield genetically-modified agricultural plants**: Bioengineering Research Center of the Russian Academy of Sciences, Moscow, 150 million rubles

4. **Development and initiation of production of promising photoelectronic matrix modules to facilitate creation of competitive Russian infrared technology**: Orion Research and Production Association, Moscow, 300 million rubles

5. **Development and launch of industrial production of new-generation catalyzers and catalytic technology for the production of engine fuel**: Boreskov Institute of Catalysis of the Russian Academy of Sciences, Novosibirsk, 350 million rubles

6. **Development and initiation of industrial production of technology for producing new types of high-quality cardboard using recycled fibers**: Central Scientific Research Institute of Paper, Pravdinsky, Moscow region, 150 million rubles

7. **Development and initiation of production of a family of highly-efficient gas-steam power units with individual generating capacity exceeding 200 MW**: Leningrad Metallic Plant, St. Petersburg, 450 million rubles

8. **Creation of technology for and initiation of industrial production of metallic construction materials with 100-percent improved utilization characteristics**: Prometei Central Scientific Research Institute of Construction Materials, St. Petersburg, 200 million rubles

9. **Development of the synthetic crystal-dielectric industry and its products**: Shubnikov Institute of Crystallography, Moscow, 460 million rubles

10. **Development and initiation of mass production of a family of competitive diesel engines for various automotive transport applications**; Zavolzhsky Engine Plant, Zavolzhye, Nizhny Novgorod Oblast, 500 million rubles

11. **Development and practical application of equipment, technologies, and organizational-financial approaches (including comprehensive measures) for improving the efficiency of the heating supply system for the regions of Russia**: Heating Investment Company, Syktyvkar, Komi Republic, 350 million rubles

12. **Improving the efficiency of the solid waste reprocessing by using modern Russian technology and equipment to produce recycled materials and commercial products**: Mekhanobrtekhnika firm, St. Petersburg, 400 million rubles

Note: Awards announced in April 2003.

Source: Russian Ministry of Industry, Science, and Technology, November 2003.

Appendix L

The Threats to Russia
(View of the Ministry for
Emergency Situations)

1. Corruption and incompetence of the governing structure
2. Increase in the hegemony of the United States
3. Increase in crime and the criminal economy
4. Lowering of the standard of living and antagonisms within the social structure
5. Decline in the production and investment potential
6. Decline in the scientific-technical and innovation potential
7. Increase in the military and technical strength of China
8. Decline in the defense and fighting capability of the military forces
9. Sharpening of the internal conflicts among nationalities and religious groups
10. Deepening of the energy crisis
11. Increase in the openness of the national economy beyond appropriate limits
12. Growth in the military threat from the United States and NATO (North Atlantic Treaty Organization)
13. Sharpening and deepening of regional and local armed conflicts
14. Increasing damage from dangerous national and catastrophic events and processes
15. Increasing damage from industrial accidents, environmental pollution, and depletion of natural resources.

Source: Kommersant (2003).

References

Ailes, C. P., and A. E. Pardee. 1984. The U.S.-USSR Agreement on Cooperation in the Fields of Science and Technology: Review and Evaluation. SRI International, Washington, D.C. March.

Alexeev, M., and L. Walker, eds. 1991. Estimating the Size of the Soviet Economy. Summary of a Meeting. Commission on Behavioral and Social Sciences and Education and Office of International Affairs. Washington, D.C.: National Academy Press.

Allakhverdov, A., and V. Pokrovsky. 2003. Academy plucks best biophysicists from a sea of mediocrity. Science (February 14): 994.

Balzer, H. 1989. Soviet Science on the Edge of Reform. Boulder, Colo.: Westview Press.

Byrnes, R. F. 1976. Soviet-American Academic Exchanges, 1958–1975. Bloomington: Indiana University Press.

Centre for Science Research and Statistics. 2003. Russian Science and Technology at a Glance, 2002. Moscow.

Dezhina, I. 1999. The International Science Foundation. New York: International Science Foundation.

Donahue, Thomas M., ed. 1991. Planetary Sciences, American and Soviet Research. Proceedings from the US-USSR Workshop on Planetary Sciences, January 2–6, 1989. Academy of Sciences of the USSR and National Academy of Sciences. Washington, D.C.: National Academy Press.

Garrett, T. 1988. Soviet Science and Technology—A New Era? Department of Trade and Industry, London. February.

Gokhberg, L. 1997. Russian R&D: A Sectoral Analysis. Science Policy Research Unit, University of Sussex. July.

Gokhberg, L., and I. Kutnetsova. 2001. Technological Innovation in Industry and Services. Centre for Science Research and Statistics, Moscow.

Graham, L. 1993. Science in Russia and the Soviet Union: A Short History. Cambridge, UK: Cambridge University Press.

Holloway, D. 1983. The Soviet Union and the Arms Race. New Haven: Yale University Press.

Institute of Ethnology and Anthropology. 2002. A Short Annotated Bibliography of the Chechen Conflict. Moscow: Institute of Ethnology and Anthropology (in Russian).

———. 2003. Peace in Chechnya through Education. Material from an International Workshop. Moscow: Institute of Ethnology and Anthropology (in Russian).

Integration of Science and Higher Education in Russia. 2001. Proceedings of the All Russian Conference with International Cooperation. Vols. 1 and 2. September 14–17. Samara-Kazan, Russia (in Russian).

International Science and Technology Center (ISTC). 2003. Annual Report for 2002. Moscow: ISTC.

Josephson, P. R. 1997. New Atlantis Revised: Akademgorodok, the Siberian City of Science. Princeton, N.J.: Princeton University Press.

Joyce, J. M. 1982. U.S.-Soviet Science Exchanges: A Foot in the Soviet Door. Russian Research Center, Harvard University, Cambridge, Mass.

Kaiser, R. G. 1988/89. The USSR in decline. Foreign Affairs (winter).

Kershenbaum, V. T., ed. 1996. Management of Technology: Russian-American Workshop. Moscow: Nauka and Tekhnika Publishing Center.

Kommersant. 2003. What threatens Russia. (March 27): 6 (in Russian).

Korchagin, A. D., and N. S. Orlov. 2001. About the Rights of the Government for the Results of Intellectual Activity. Experiences of the United States and the Meeting of Russian Needs. Patents and Licenses, M, No. 7. Russian Patent Office, Moscow (in Russian).

Legvold, R. 2002–2003. All the way, creating the U.S.-Russian alliance. National Interest (winter): 21–31.

Lewin, Walter H. G., George W. Clark, and Rashid Sunyaev, eds. 1991. High-Energy Astrophysics: American and Soviet Perspectives. Proceedings from the US-USSR Workshop on High Energy Astrophysics, June 18–July 1, 1989. Academy of Sciences of the USSR and National Academy of Sciences. Washington, D.C.: National Academy Press.

Marchuk, G. I. 1995. Meeting and Reflections. Moscow: Mir Publishing House (in Russian).

Martyushov, Y. S. 2003. The normative legal basis for innovation. Innovation Economy of Russia. December–February (in Russian).

McKinsey Global Institute. 1999. Unlocking Economic Growth in Russia. Moscow. October.

Medvedev, Z. 1978. Soviet Science. New York: Norton.

National Academy of Sciences (NAS). 1977. A Review of U.S.-U.S.S.R. Inter-academy Exchanges and Relations (Kaysen Panel). PB276719/AS. Springfield, Va.: National Technical Information Service.

National Academy of Sciences, Institute of Medicine, National Research Council (NAS/IOM/NRC). 1997. Controlling Dangerous Pathogens, a Blueprint for U.S. Russian Cooperation. A Report to the Cooperative Threat Reduction Program of the U.S. Department of Defense. October, Washington, D.C.

National Research Council (NRC). 1980. NAS-ASUSSR activities curtailed. Newsletter of the Soviet-East European Exchange Program of the National Academy of Sciences. Washington, D.C.

————. 1982. Scientific Communication and National Security. Panel on Scientific Communication and National Security, Committee on Science, Engineering, and Public Policy. Washington, D.C.: National Academy Press.

————. 1982–1983. Paleomagnetic and Biostratigraphic Expedition in Siberia. Newsletter of the Soviet-East European Exchange Program of the National Academy of Sciences (winter). Washington, D.C.

————. 1983–1984a. MacArthur Foundation supports Academy's USSR program. Newsletter of the Soviet-East European Exchange Program of the National Academy of Sciences (winter). Washington, D.C.

————. 1983–1984b. U.S. planetary scientist meets with Soviet colleagues. Newsletter of the Soviet-East European Exchange Program of the National Academy of Sciences (winter). Washington, D.C.

————. 1985a. NAS delegation holds talks with Soviet Academy. Newsletter of the Soviet-East European Exchange Program of the National Academy of Sciences (spring). Washington, D.C.

————. 1985b. Proof of Bieberbach conjecture verified in Leningrad. Newsletter of the Soviet-East European Exchange Program of the National Academy of Sciences (spring). Washington, D.C.

————. 1986a. NAS hosts Soviet energy specialists. Newsletter, Soviet-East European Program of the National Academy of Sciences/National Research Council (winter). Washington, D.C.

————. 1986b. Newsletter, Soviet-East European Program of the National Academy of Sciences/National Research Council (winter). Washington, D.C.

————. 1987a. American geologists discover new mineral localities in Russia. Newsletter, Soviet-East European Program of the National Academy of Sciences/National Research Council (winter). Washington, D.C.

————. 1987b. Balancing the National Interest: U.S. National Security Export Controls and Global Economic Competition. Panel on the Impact of National Security Controls on International Technology Transfer, Committee on Science, Engineering. and Public Policy. Washington, D.C.: National Academy Press.

————. 1987c. Cooperation begins with Soviet Academy on social science research and nuclear war. Newsletter, Soviet-East European Program of the National Academy of Sciences/National Research Council (summer). Washington, D.C.

————. 1987d. Evaluations of US-USSR bilateral cooperation. Newsletter, Soviet-East European Program of the National Academy of Sciences/National Research Council (summer). Washington, D.C.

————. 1988a. Baykal and Rio Grande Rift Studies. Newsletter, Soviet-East European Program of the National Academy of Sciences/National Research Council (summer). Washington, D.C.

———. 1988b. First agreement between IOM and Soviet Medical Academy. Newsletter, Soviet-East European Program of the National Academy of Sciences/National Research Council (summer). Washington, D.C.

———. 1988c. NAS vice president travels to Novosibirsk, Newsletter, Soviet-East European Program of the National Academy of Sciences/National Research Council (summer). Washington, D.C.

———. 1988d. Newsletter, Soviet-East European Program of the National Academy of Sciences/National Research Council (summer). Washington, D.C.

———. 1988e. Soviet academicians visit NAS during Washington summit. Newsletter, Soviet-East European Program of the National Academy of Sciences/National Research Council (summer). Washington, D.C.

———. 1988f. Soviet plants added to Alaska collection. Newsletter, Soviet-East European Program of the National Academy of Sciences/National Research Council (summer). Washington, D.C.

———. 1988g. US-USSR cooperation in nuclear reactor safety. Newsletter, Soviet-East European Program of the National Academy of Sciences/National Research Council (summer). Washington, D.C.

———. 1989a. American economists witness perestroika first hand. Newsletter, Soviet-East European Program of the National Academy of Sciences/National Research Council (spring). Washington, D.C.

———. 1989b. Energy specialists meet in Yalta. Newsletter, Soviet-East European Program of the National Academy of Sciences/National Research Council (spring). Washington, D.C.

———. 1989c. NAS assists in sending team of earthquake experts to Armenia. Newsletter, Soviet-East European Program of the National Academy of Sciences/National Research Council (spring). Washington, D.C.

———. 1989d. NAS hosts seminar on energy and dual use technology. Newsletter, Soviet-East European Program of the National Academy of Sciences/National Research Council (spring). Washington, D.C.

———. 1989–1990a. Economics dialogue continues with Soviet Academy. Newsletter, Soviet-East European Program of the National Academy of Sciences/National Research Council (winter). Washington, D.C.

———. 1989–1990b. Energy and ecology specialists visit Siberia. Newsletter, Soviet-East European Program of the National Academy of Sciences/National Research Council (winter). Washington, D.C.

———. 1989–1990c. Future development and safety of nuclear power. Newsletter, Soviet-East European Program of the National Academy of Sciences/National Research Council (winter). Washington, D.C.

———. 1990a. Gorbachev advisers visit NAS. Newsletter, Soviet-East European Program of the National Academy of Sciences/National Research Council (fall). Washington, D.C.

———. 1990b. Newsletter, Soviet-East European Program of the National Academy of Sciences/National Research Council (fall). Washington, D.C.

———. 1990c. Radioactive waste seminar in the USSR. Newsletter, Soviet-East European Program of the National Academy of Sciences/National Research Council (fall). Washington, D.C.

———. 1990d. Soviet-American Dialogue in the Social Sciences, Research Workshops on Interdependence among Nations. Committee on Contributions of Behavioral and Social Science to the Prevention of Nuclear War, Commission on Behavioral and Social Sciences and Education. Washington, D.C.: National Academy Press.

———. 1990e. Soviet Social Science: The Challenge for the American Academic Community. Summary of a Meeting. Commission on Behavioral and Social Sciences and Education. Washington, D.C.: National Academy Press.

———. 1990f. Young U.S. and Soviet economists meet in Boston. Newsletter, Soviet-East European Program of the National Academy of Sciences/National Research Council (fall). Washington, D.C.

———. 1991a. Dramatic changes in USSR and Eastern Europe lead to new approaches in exchanges in 1993. Newsletter, Soviet-East European Program of the National Academy of Sciences/National Research Council (fall). Washington, D.C.

———. 1991b. Finding Common Ground: U.S. Export Controls in a Changed Global Environment. Panel on the Future Design and Implementation of U.S. National Security Export Controls, Committee on Science, Engineering, and Public Policy. Washington, D.C.: National Academy Press.

———. 1991c. Summer programs for young investigators in economics, cosmology, and string theory. Newsletter, Soviet-East European Program of the National Academy of Sciences/National Research Council (fall). Washington, D.C.

———. 1992a. Improving Social Science in the Former Soviet Union. Commission on Behavioral and Social Sciences and Education. Washington, D.C.: National Academy Press.

———. 1992b. NAS and Europeans discuss science in FSU. Newsletter, Office for Central Europe and Eurasia of the National Academy of Sciences/National Research Council. Washington, D.C.

———. 1992c. NAS opens science and technology policy dialogue with the Russian Academy. Newsletter, Office for Central Europe and Eurasia of the National Academy of Sciences/National Research Council. Washington, D.C.

———. 1992d. Reorientation of the Research Capability of the Former Soviet Union. National Academy of Sciences, National Academy of Engineering, Institute of Medicine, Washington, D.C.

———. 1993a. NAS-RAS working group on new forms of cooperation. Newsletter, Office for Central Europe and Eurasia of the National Academy of Sciences/National Research Council. Washington, D.C.

———. 1993b. 1993 Young Investigator Programs. Newsletter, Office for Central Europe and Eurasia of the National Academy of Sciences/National Research Council. Washington, D.C.

———. 1993c. Redeploying Assets of the Russian Defense Sector to the Civilian Economy. Final Report of the Committee on Enterprise Management in a Market Economy under Defense Conversion. Office of International Affairs. Washington, D.C.: National Academy Press.

———. 1993d. A Russian-American Partnership for Industrial Development (RAPID). Committee on Enterprise Management in a Market Economy under Defense Conversion, Office of International Affairs. Washington, D.C.: National Academy Press.

———. 1993e. Sustaining Excellence in Science and Engineering in the Former Soviet Union. National Academy of Sciences, National Academy of Engineering, Institute of Medicine, Washington, D.C.: National Academy Press.

———. 1993f. Young Investigator Program on Ecological Concerns in the Development of the Arctic and Far Northern Regions of Russia. Office for Central Europe and Eurasia. Washington, D.C.

———. 1994. Dual-Use Technologies and Export Control in the Post-Cold War Era. Documents from a Joint Program of the National Academy of Sciences and the Russian Academy of Sciences. Office of International Affairs. Washington, D.C.: National Academy Press.

———. 1995. Oil wastes contaminate Western Siberia. Newsletter, Office for Central Europe and Eurasia. Washington, D.C.

———. 1996a. An Assessment of the International Science and Technology Center. Office of International Affairs. Washington, D.C.: National Academy Press.

———. 1996b. Cooperation in applied science and technology. Newsletter, Office for Central Europe and Eurasia. Washington, D.C.

———. 1996c. Young Investigator Program on Sustainable Forestry, Group Scientific Report. Office for Central Europe and Eurasia. Washington, D.C.

———. 1996d. Young Investigator Program on Water Quality, Group Scientific Report. Office for Central Europe and Eurasia. Washington, D.C.

———. 1997a. Proliferation Concerns, Assessing U.S. Efforts to Help Contain Nuclear and Other Dangerous Materials and Technologies in the Former Soviet Union. Office of International Affairs. Washington, D.C.: National Academy Press.

———. 1997b. Russian forest specialists visit the Pacific Northwest. Newsletter, Office for Central Europe and Eurasia. Washington, D.C.

———. 1997c. Russians and Americans cooperate on protecting water resources. Newsletter, Office for Central Europe and Eurasia. Washington, D.C.

———. 1997d. A successful conclusion to the CAST program. Newsletter, Office for Central Europe and Eurasia. Washington, D.C.

———. 1998a. Partners on the Frontier, U.S.-Russian Cooperation in Science and Technology. Proceedings of a Workshop. Office for Central Europe and Eurasia. Washington, D.C.: National Academy Press.

———. 1998b. Technology Commercialization: Russian Challenges, American Lessons. Committee on Utilization of Technologies Developed at Russian Research and Educational Institutions, Office of International Affairs. Washington, D.C.: National Academy Press.

———. 1999. Protecting Nuclear Weapons Materials in Russia. Committee on Upgrading Russian Capabilities to Secure Plutonium and Highly Enriched Uranium, Office of International Affairs. Washington, D.C.: National Academy Press.

———. 2001. The Role of Environmental NGOs—Russian Challenges, American Lessons: Proceedings of a Workshop. Office for Central Europe and Eurasia, Development, Security, and Cooperation, Policy and Global Affairs. Washington, D.C.: National Academy Press.

———. 2002a. Current visa restrictions interfere with U.S. science and engineering contributions to important national needs. Statement of the three presidents of the National Academies. Washington, D.C. December 13.

———. 2002b. High Impact Terrorism: Proceedings of a Russian-American Workshop. Committee on Confronting Terrorism in Russia, Office for Central Europe and Eurasia. Washington, D.C.: National Academy Press.

———. 2002c. Successes and Difficulties of Small Innovative Firms in Russian Nuclear Cities: Proceedings of a Russian-American Workshop. Committee on Small Innovative Firms in Russian Nuclear Cities, Office for Central Europe and Eurasia. Washington, D.C.: National Academy Press.

———. 2003a. Conflict and Reconstruction in Multiethnic Societies: Proceedings of a Russian-American Workshop. Washington, D.C.: The National Academies Press.

———. 2003b. End Points for Spent Nuclear Fuel and High Level Radioactive Waste in Russia and the United States. Board on Radioactive Waste Management. Washington, D.C.: The National Academies Press.

———. 2003c. Letter Report from the Co-Chairs of the Joint Committee on U.S.-Russian Cooperation on Counterterrorism Challenges for Russia and the United States. June.

Office of Science and Technology Policy (OSTP). 1985. A Study of Soviet Science. Washington, D.C.: Government Printing Office.

Office of the Undersecretary of Defense for Policy. 1985. Assessing the Effect of Technology Transfer on U.S./Western Security: A Defense Perspective. Washington, D.C. February.

Organization for Economic Cooperation and Development (OECD). 1994. Cooperation in Science and Technology with the Federation of Russia: Experience with Programs of Selected OECD Countries. Paris.

———. 2001. Bridging the Innovation Gap in Russia: Science and Innovation. The Helsinki Seminar. Paris.

Popova, T. 2002. Nord Ost through the Eyes of a Hostage. Moscow: Vagrius Publishing Co. (in Russian).

Richmond, Y. 2003. Cultural Exchange and the Cold War: Raising the Iron Curtain. University Park: Penn State University Press.

Rossiskaya Akademiya Nauk 1991–2001. 2002. Moscow: Nauka Publishing Co. (in Russian).

Russian Academy of Sciences (RAS). 2002. Rating Higher Education Universities for 2001. Poisk. May 17.

Sagdeev, R. Z. 1994. The Making of a Soviet Scientist, My Adventures in Nuclear Fusion and Space from Stalin to Star Wars. New York: John Wiley.

Sakharov, A. 1990. Memoirs. New York: Knopf.

Schweitzer, G. E. 1988. Who wins in U.S.-Soviet science ventures? Bulletin of the Atomic Scientists (October).

———. 1989. Techno-diplomacy: US-Soviet Confrontations in Science and Technology. New York: Plenum Press.

———. 1992. U.S.-Soviet scientific cooperation. Technology in Society 14: 173–185.

———. 1996. Moscow DMZ: The Story of the International Effort to Convert Russian Weapons Science to Peaceful Purposes. Armonk, N.Y.: M. E. Sharpe.

———. 1997. Experiments in Cooperation: Assessing U.S.-Russian Programs in Science and Technology. New York: Twentieth Century Fund.

———. 2000. Swords into Market Shares: Economic, Security, and Technology Challenges in the New Russia. Washington, D.C.: Joseph Henry Press.

———. 2001. Industrial Perestroika, High Tech, Innovation, and Economic Growth in Russia. Arlington, Va.: Cameron Publication Services.

Statute of the Russian Academy of Sciences, 1724–1999. 1999. Moscow: Nauka Publishing Co.

Thursby, J. G., and M. C. Thursby. 2003. University licensing and the Bayh Dole Act. Science (August 22): 1052.

United Nations Development Programme (UNDP). 2003. UN Common Country Assessment for the Russian Federation, 2002. Moscow.

U.S. Congress, Commission on Security and Cooperation in Europe. 1988. Reform and Human Rights: The Gorbachev Record. Washington, D.C. May.

U.S. Congress, Joint Economic Committee. 1987. Gorbachev's Economic Plans. Vols. 1 and 2. Washington, D.C.

U.S. Department of State. 2003. U.S. Government Assistance to and Cooperative Activities with Eurasia, FY 2002 Annual Report. Office of Coordination of U.S. Assistance to Europe and Eurasia. March.

U.S. House of Representatives. 1986. Hearings of the Subcommittee on Europe and the Middle East. Committee on Foreign Affairs. July 31. Washington, D.C.

Vucinich, A. 1984. Empire of Knowledge: The Academy of Sciences of the USSR (1917–1980). Berkeley: University of California Press.

Walker, L., and P. Stern. 1993. Balancing and Sharing Political Power in Multiethnic Societies: Summary of a Workshop. Washington, D.C.: National Academies Press.

Watkins, A., S. Bossourtrot, and L. Poznanskaya. 2001. Russian science and technology for a market economy. Unpublished manuscript released by World Bank, Washington, D.C.

Woodrow Wilson Center and Smithsonian Institution. 1984. Exchanges: Ripoff or Payoff? Washington, D.C. November.

World Bank. 2003. Russian Economic Report #6: August 2003. Online. Available at www.worldbank.org.ru/.

World Health Organization (WHO). 2003. WHO Country Cooperation Strategy, Russian Federation. Geneva.